T0312686

Integrating Work Health and Safety into Construction Project Management

Integrating Work Health and Safety into Construction Project Management

Helen Lingard
RMIT Distinguished Professor, School of Property, Construction and Project Management, RMIT University, Melbourne, Australia

Ron Wakefield
Professor and Dean, School of Property, Construction and Project Management, RMIT University, Melbourne, Australia

This edition first published 2019
© 2019 John Wiley & Sons Ltd

Registered Offices
John Wiley & Sons, Inc., 111 River Street, Hoboken, NJ 07030, USA
John Wiley & Sons Ltd, The Atrium, Southern Gate, Chichester, West Sussex, PO19 8SQ, UK

Editorial Office
9600 Garsington Road, Oxford, OX4 2DQ, UK

For details of our global editorial offices, customer services, and more information about Wiley products visit us at www.wiley.com.

Wiley also publishes its books in a variety of electronic formats and by print-on-demand. Some content that appears in standard print versions of this book may not be available in other formats.

Library of Congress Cataloging-in-Publication Data

Names: Lingard, Helen, author. | Wakefield, Ron author.
Title: Integrating Work Health and Safety into Construction Project
  Management / Helen Lingard, RMIT Distinguished Professor, School of
  Property, Construction and Project Management, RMIT University,
  Melbourne, Australia, Ron Wakefield, Professor and Dean, School of
  Property, Construction and Project Management, RMIT University, Melbourne,
  Australia.
Description: Hoboken, NJ, USA : Wiley-Blackwell, [2019] | Includes
  bibliographical references and index. |
Identifiers: LCCN 2019000828 (print) | LCCN 2019001355 (ebook) | ISBN
  9781119159940 (AdobePDF) | ISBN 9781119159957 (ePub) | ISBN 9781119159926
  (hardcover)
Subjects: LCSH: Building–Safety measures.
Classification: LCC TH443 (ebook) | LCC TH443 .L56 2019 (print) | DDC
  624.068/3–dc23
LC record available at https://lccn.loc.gov/2019000828

Cover Design: Wiley
Cover Image: © Sirisak Boakaew/Getty Images

Set in 10/12pt WarnockPro by SPi Global, Chennai, India
Printed in Singapore by Markono Print Media Pte Ltd

10 9 8 7 6 5 4 3 2 1

Contents

Preface

This book presents a synthesis of more than a decade of research conducted by a small, multidisciplinary team of researchers at RMIT University in Melbourne, Australia. The idea for the book grew from our reflections about lessons learned from the research and, in particular, the way our own thoughts about work health and safety (WHS) in the construction industry have developed and changed over time.

From the outset, our collaborative research activity was driven by a shared belief that something different needed to be done to prevent the relatively high incidence of work-related death, injury, and illness experienced by construction workers. Together with Professor Nick Blismas (a significant contributor to several chapters in this book), Helen Lingard and Ron Wakefield initiated a programme of research to better understand and directly address the barriers to improving WHS in the construction industry. Our earliest work, undertaken at the Tullamarine-Calder Interchange Alliance, was strongly supported by Pat Cashin, General Manager of Baulderstone Pty Ltd. This work grew into a multipronged programme, involving many different partner organizations and guided by an active Industry Advisory Group chaired by former National President of Engineers Australia, Peter Godfrey.

The backdrop to the programme of research presented in this book was a growing international focus on the role to be played by clients and designers in identifying and addressing WHS risks in their decision making. Prompted by the recognition that some WHS risks experienced by construction workers could be traced back to planning and design choices, the RMIT research team was engaged by the Cooperative Research Centre for Construction Innovation to develop a voluntary Guide to Best Practice for Safer Construction. The Guide was commissioned by Engineers Australia and its development was led by an industry task force consisting of peak bodies representing contractors, design consultants, and public and private sector construction clients. The Guide established a set of principles to drive collaboration and sharing of WHS responsibility between clients, designers, and constructors, and also suggested WHS management practices for each stage in the project lifecycle, from planning and design through to construction and completion.

However, the Guide did not reflect the social and technical complexity of construction projects. It treated WHS as something that could be managed through a mechanistic process of risk identification, assessment, and control within each project stage. The

Guide established simplistic roles and responsibilities for clients, designers, and constructors without acknowledging the heterogeneous nature of client organizations, or the complex web of designers and constructors involved in project delivery. Neither did the Guide adequately reflect the fact that construction project participants' actions and decisions are shaped by broader forces in regulatory, economic, and policy contexts.

Our understanding of the factors at play in shaping client behaviour and safety in design effectiveness became more nuanced as we considered the impact of organizational complexity, procurement policy, supply network fragmentation, and the segregation of product and process design. Our work also expanded into new areas. We investigated how aspects of an organizational culture impact WHS, and we considered how workers' health and wellbeing are shaped by the work practices and the quality of work in the construction industry.

In 2009, Helen was awarded an Australian Research Council Future Fellowship to undertake a four-year programme of work investigating the importance of integration to protecting construction workers' WHS. The Future Fellowship programme of work, titled 'Differentiation not disintegration: Integrating strategies to improve occupational health and safety in the construction industry' (ID number FT0990337), has provided a strong backbone for this book.

While each chapter of this book can be read as a standalone presentation of our work on a particular topic, we encourage readers to explore and reflect on the points of connection between the information contained in different chapters. For example, the issue of workers' health and wellbeing cannot be properly understood without considering the timelines established for delivering projects and the implications of tight project schedules for hours of work, the quality of work–family interactions, and wellness in the workforce.

In writing this book our overarching aim was to explore many topics in construction WHS through the theme of integration. We suggest WHS needs to be an integral part of managing construction organizations and projects, such that it constitutes a serious consideration in everything that is done.

We do not favour glib statements that WHS should be an organization's 'number one priority'. Indeed, such statements are cynically received when workers are fully aware that managers are rewarded for performance on multiple competing priorities. However, WHS does need to be firmly embedded in decisions made about all aspects of business and project management. WHS should not be treated as an afterthought, to be considered once important decisions have already been made. Unfortunately, managerial decisions with the potential to impact WHS are sometimes post-rationalized, with the result that the most effective forms of risk control are not realized.

It is also vital that we remain alert to the main aim of managing WHS, which is to protect the health and safety of workers. In this book we sought to provide insights gleaned from our research to suggest ways to more effectively integrate WHS into management decision making for the purpose of making workplaces and systems of work safer and healthier.

The book represents a team effort. Our colleagues have played a key role in working on specific topics or research projects. These team members have significantly contributed to the development of our thinking and, in this book, they are acknowledged as co-authors of chapters about topics they have worked on. We extend warm and very grateful thanks to these colleagues, whose ideas, passion, and hard work have greatly enriched the collective research programme.

We also gratefully acknowledge the important groundwork provided by researchers whose formative contributions helped shape the research presented in the book, in particular, Tracy-Lee Cooke and David Jellie.

Last, but certainly not least, we acknowledge the support of organizations and agencies that have funded components of the research, including (in alphabetical order):

- Australian Constructors Association
- Australian Research Council
- Baulderstone Pty Ltd
- CodeSafe Solutions
- Department of Justice and Attorney General, Queensland Government
- Department of State Development Business and Innovation, Victorian Government
- Fonterra Cooperative Group Pty Ltd
- Lendlease
- Major Projects Victoria
- Major Transport Infrastructure Program, Department of Economic Development, Jobs, Transport and Resources, Victorian Government.
- National Institute for Occupational Safety and Health (NIOSH) (under a subcontract agreement with Virginia Tech.)
- Office of the Federal Safety Commissioner
- Port of Melbourne Corporation
- Probuild
- Treasury and Economic Development Directorate, Australian Capital Territory Government
- WorkSafe Victoria.

Helen Lingard and
Ron Wakefield

1

The State of Work Health and Safety in Construction

1.1   The Construction Safety Problem

Most reports or articles about work health and safety (WHS) in construction begin with a statement about the industry's poor safety statistics. Irrespective of the part of the world in which a particular study has been conducted, it is common for authors to describe:

- high rates of injury and fatality in construction, relative to other industries, and
- disproportionate numbers of work-related injuries or deaths compared to the size of the construction workforce.

The construction WHS problem is a global one. Indeed, the International Labour Organization (ILO) estimates at least 60 000 fatal accidents occur in construction each year, representing one fatal accident every 10 minutes. The ILO estimates the construction sector typically employs between 6% and 10% of the workforce, but accounts for between 25% and 40% of work-related deaths.

The Center for the Protection of Workers' Rights Construction Chart Book (2013) provides information about the leading causes of work-related fatalities and non-fatal work injuries resulting in days away from work (DAFW) in the construction industry in the USA. Between 1992 and 2010, the highest ranked causes of fatalities in construction were:

- falls to a lower level (6678 deaths);
- highway incidents (2707 deaths);
- contact with electric current (2443 deaths); and
- being struck by an object (2054 deaths).

In contrast, there were 74 950 reported non-fatal injuries resulting in DAFW in the USA construction industry in 2010. Leading causes were:

- bodily reaction/exertion (33.6%);
- contact with objects (33.0%); and
- falls (24.2%).

Integrating Work Health and Safety into Construction Project Management, First Edition.
Helen Lingard and Ron Wakefield.
© 2019 John Wiley & Sons Ltd. Published 2019 by John Wiley & Sons Ltd.

In Australia, an industry profile compiled in 2018 found the most common types of incidents resulting in serious claims for workers' compensation between 2012–2013 and 2015–2016 were:

- muscular stress while lifting, carrying, or putting down objects (16%);
- muscular stress while handling objects (14%);
- falls on the same level (13%);
- falls from height (11%);
- being hit by a moving, or flying object (8%);
- hitting moving objects (6%); and
- other mechanisms (32%) (Safe Work Australia 2018).

Of the construction fatalities that occurred in Australia between 2013 and 2016, the majority involved:

- falls from height (30%);
- being hit by falling objects (15%);
- vehicle incidents (15%);
- being hit by moving objects (11%);
- contact with electricity (10%);
- being trapped between stationary and moving objects (9%); and
- other mechanisms (11%) (Safe Work Australia 2018).

The largest number of fatalities involved construction and mining labourers (22% or 27% fatalities over the four-year period). Other occupations involved in fatalities were electricians (11% or 14% fatalities), bricklayers, carpenters, and joiners (8% or 10% fatalities), and mobile plant operators (8% or 10% fatalities) (Safe Work Australia 2018).

In the UK, there were 196 fatal injuries to workers in the construction sector between 2012–2013 and 2016–2017. Of these:

- 97 involved a fall from height;
- 19 involved someone being trapped by something collapsing or overturning;
- 19 involved someone being struck by a moving vehicle;
- 16 involved someone being struck by a moving, including flying, object;
- 14 involved contact with electricity or an electrical discharge; and
- 9 involved contact with moving machinery (Health and Safety Executive 2018a).

Non-fatal injuries to construction workers in the UK in 2016–2017 that resulted in more than seven days off work involved:

- lifting/handling (29%);
- slips, trips, or falls on the same level (21%);
- falls from height (10%);
- struck by moving, including flying, object (12%);
- contact with moving machinery (6%); and
- struck by moving vehicle (1%) (Health and Safety Executive 2018a).

The evidence suggests safety performance of construction industries in developing countries is considerably poorer than in developed countries. This may be because institutional and governance frameworks regulating industrial activities are relatively weak and have little impact (Kheni et al. 2008) and because the construction industry in developing countries relies on an unskilled, mobile workforce, often drawn from agricultural backgrounds (Priyadarshani et al. 2013). The economic environment in many developing countries also creates challenges for WHS as construction businesses operate in a competitive, relatively unregulated, environment. Delayed payments, and the failure of contractor assistance programmes, dramatically reduce resources available for investment in improving workers' health and safety (Kheni et al. 2010).

In the USA, Australia, and the UK, recent decades have seen a steady downward trend in rates of non-fatal injury in the construction industry. In contrast, projections for developing countries are for an increase in work-related injuries and deaths as work becomes more industrialized (Kheni et al. 2008). In the UK, Australia, and the USA the numbers of work-related fatalities in construction have also declined, although the rate of fatalities remains high relative to other industries. In the UK, the fatality rate of 1.62 per 100 000 workers per year is more than 3.5 times the average rate across all industries (0.46 per 100 000 workers) (Health and Safety Executive 2017). The Center for the Protection of Workers' Rights observes that reductions in fatalities have not occurred uniformly across all incident types. Thus, in the USA, fatalities due to contact with electric current decreased nearly 45% between 1995 and 2010, while the number of fatalities from falls to a lower level was similar at the two time points. Also, the total number of deaths due to highway incidents became the second leading cause of fatalities in construction over the period 1995–2010. Although deaths in some areas have reduced, in others they have remained fairly constant (CPWR 2013).

A detailed comparative analysis of international safety statistics is beyond the scope of this introductory chapter. However, the quick overview of statistics from the USA, Australia, and the UK reveals some important insights for preventing work-related injury and fatalities.

First, the ways in which construction workers are injured and killed (at least in industrialized countries) are remarkably similar and have changed little over recent years. The same injury mechanisms and incident classifications are prevalent, meaning construction workers are still being killed and injured in ways that are well-known and documented in national and international statistical reports.

Second, although work-related injuries have decreased in many countries, on average the construction industry's fatality rate remains relatively high, and some types of incident have been resistant to change.

Third, the type of incident that results in a non-fatal injury (albeit one that involves a workers' compensation claim) is generally quite different from the type of incident in which someone is killed.

The implications of these three observations will be considered briefly in turn.

The similarity between injuries and incident types, over time and across the globe, indicate that the kinds of activities and incidents that result in people being injured or killed are known and understood. Further, there is not a great deal of variation between these activities and incidents among construction industries (at least in industrialized countries). Fatal incidents are largely attributed to falls from height, contact with electricity, and being trapped or struck by a moving object. Body exertion, lifting/handling, falling, and being struck by an object were leading incident types resulting in non-fatal injury. The consistency with which these types of incidents/injuries impact on construction workers indicates that strategies targeting these specific areas could significantly reduce the burden of injury or death in the construction industry.

Data from the USA suggest some types of fatal incidents have been reduced through targeted collective industry efforts. Most notably, the number of fatal incidents involving workers coming into contact with electricity reduced in recent years. However, in other areas, including falls and highway incidents, fatalities have not reduced to the same extent. The persistence of certain types of fatal incident suggests greater efforts need to be targeted to reducing work-related deaths in these high-risk, high-consequence areas.

Finally, differences in the types of incident that produce low- versus high-consequence outcomes can have implications for where resources and effort are focused.

Some writers on WHS have suggested a false sense of invulnerability in high-risk organizational environments has resulted from the emphasis on lost time injury frequency rates, and consequent effort focused on preventing occupational injuries seen as being high in frequency but of low consequence. To evidence this argument, it is pointed out that serious incidents resulting in multiple fatalities, as well as extensive environmental damage and service disruption, have occurred in organizations believed to have good safety records, based on the measurement of occupational injury frequency rates. Two often-cited examples are an explosion at the Longford gas facility in Australia (Hopkins 2000), and the blow out, subsequent explosion, and uncontrollable fire at the Macondo (Deepwater Horizon) well in the Gulf of Mexico (Dekker 2014).

It is argued that effective control of occupational injury frequency rates at these sites – Dekker describes how the managers of the Deepwater Horizon well had prohibited carrying coffee in a cup without a lid – masked an underlying, gradual, and incremental drift towards failure danger as the production systems at Longford and Macondo edged closer to the edge of their safety 'envelope'. The argument has also been made that indicators of occupational safety performance are not good measures of how effectively process safety risks are being controlled (see, for example, Baker 2007). The point is often made by people studying high-risk production processes, such as those found in the oil and gas or nuclear energy industries, that unlike the majority of occupational safety risks, process safety risks have the potential to cause harm to workers and the general public on a very large scale. While these arguments have some validity, taking this thinking to its logical conclusion in an industry such as construction may not be helpful. Many work-related injuries and illnesses experienced by construction workers are very high in frequency, yet have non-fatal consequences (for example musculoskeletal disorders). These injuries and illnesses cause significant pain, disability, and hardship for workers. They need to be the focus of concerted prevention efforts (see also Chapter 8) at the same time as managing risks associated with high-consequence failures.

The question has also been raised about whether safety incidents producing outcomes with different degrees of severity share similar causes. The 'similar causation' argument stems from work undertaken by Heinrich (1931) who investigated several thousands of insurance claims for deaths and disabling injuries. Heinrich studied the history of activities being undertaken when these incidents occurred and collated statistics showing the relative frequency for these activities of serious/disabling injury, minor injury, and near-miss incidents. He found that for every serious/disabling injury, there were many more minor injuries and more near misses again. Hale (2002) describes how, as a result of Heinrich's analysis, it has become an 'urban myth' that incidents resulting in serious injury share the same causes as those resulting in minor injury

(Hale, 2002). Hale (2002) observes that Heinrich's analysis of causation was never made clear. Because not all minor incidents could have been major incidents, it is largely due to careless reasoning that safety practitioners have come to expect that preventing minor incidents (low consequence) will automatically lead to preventing major incidents resulting in death or permanent disability. The amount of damage that occurs is, according to Hale (2002), a factor of the amount of damaging energy that is released in a particular situation, and what it comes into contact with before it dissipates. High-energy activities and events will largely produce more damage, and more serious consequences, than low-energy activities and events (see also Hallowell et al. 2017).

Bellamy (2015) re-examined this argument, modelling the causes of 23 000 reportable fatal and non-fatal incidents occurring in the Netherlands between 1998 and 2003. This study revealed that incident causes were similar for fatal and non-fatal incidents, but only if looking within the same hazard category (or incident type). The analysis also reveals that, although incidents of a similar type share similar causes, these causes were not observed in the same proportions. Thus, causes relevant to fatal falls from height (roofs/platforms/floors) were:

- fall-arrest failure (48% of fatal incidents) and
- roof edge-protection failure (42% of fatal incidents).

Roof edge-protection failure was a factor in 45% of non-fatal falls from roofs/platforms or floors, but fall-arrest failure was a factor in only 28% of non-fatal incidents of this type (Bellamy 2015).

Bellamy (2015) observes a similar finding for falls from a scaffold. Edge-protection failure was a factor in 44% of fatal, and 31% of non-fatal, falls from a scaffold. Deficient anchoring or fixings was a factor in 30% of fatal, and 20% of non-fatal, falls from scaffolds. Loss of control of body balance was involved in proportionally more non-fatal falls from a scaffold (39%), compared to fatal falls from a scaffold (26%).

On the basis of these findings, Bellamy (2015) suggests the analysis of minor (non-fatal) occupational accidents can help to prevent major (fatal) ones, providing incidents of the same hazard or type are analysed together.

The international construction safety statistics support the assertion that different hazards or incident types have different degrees of lethality. Some types of hazard are far more likely to result in serious consequences (such as a fatality) than others. However, it is important not to lose sight of some of the high-frequency, low- (or lower-) consequence WHS issues that impact construction workers as these create considerable human cost and social and economic impact. Hale (2002) concludes: 'We should discriminate between the scenarios that can lead to major disaster and those which can never get further than minor inconvenience. If we tackle minor injury scenarios, it should be because minor injuries are painful and costly enough to prevent in their own right, not because we believe the actions might control major hazards' (p. 40).

Trade unions have also noted the importance of managing risks associated with frequent but relatively low-consequence incidents. In the UK, the Union of Construction, Allied Trades and Technicians stated that: 'Small injuries can mean significant loss of pay and significant psychological stress for the worker and their family. If we don't have zero tolerance in the work place, then standards will slip and the number of injuries will increase … Allowing workers to suffer small injuries, while focusing just on saving lives, is not good for building workers. Building workers need to stay completely and entirely safe – and avoid all injuries' (Warburton 2016).

1.2   The Neglect of Occupational Health

Another observation to make about construction WHS is that historically, a strong emphasis has been placed on safety. However, much less attention has been paid to problems relating to construction workers' health. This is in spite of the fact that work-related illness is a very significant problem in the construction industry, and workers are exposed to a multitude of serious occupational health hazards in their daily work.

Snashall (2005) reports that construction workers have a high overall mortality rate, independent of social class. Further, Snashall (2005) points out that, because of the diversity of construction jobs and activities, almost every occupational illness has been recorded among construction workers.

Silica is a particularly insidious occupational hazard in construction. Silica is found in sand, granite, quartz, and most stone. Fine, respirable particles of crystalline silica dust are created when these materials are chipped, cut, drilled, or ground. Exposure to silica dust causes silicosis, a disabling and often fatal disease similar to black lung experienced by coal miners (Lahiri et al. 2005).

Many construction activities involve exposure to silica dust, including:

- abrasive blasting with sand;
- jack hammering;
- rock/well drilling;
- concrete mixing;
- concrete drilling;
- brick and concrete block cutting and sawing;
- tuck pointing; and
- tunnelling operations (OSHA 2002).

Even very small amounts of silica dust can cause harm, and by the time symptoms become apparent the condition is often serious, leading to permanent disability or death. In the UK, it is estimated that every year more than 500 construction workers die from exposure to silica dust (Health and Safety Executive 2013).

Part of the problem may be in how WHS is conceptually framed. In the commonly used acronym 'WHS', workers' health and safety are typically referred to in the singular. Although this is a semantic point, it is also an important one because the evidence suggests occupational health hazards to which construction workers are exposed are being addressed less effectively than the safety hazards. Implicit in the interfusion of health and safety into a single concept is the implication that health and safety can both be managed in the same way by the same processes; that is, while focusing effort on improving workers' safety, their health will also somehow be magically improved. However, the high rates of serious work-related illness in the construction industry suggest this is not the case, and health needs to be better managed as a separate issue. Sherratt (2015) argues standard WHS risk management processes are not well suited to managing occupational health risks that require special attention and a different approach.

In the UK, the Health and Safety Executive reports that, in 2018, 82 000 construction workers suffered from a new or long-standing work-related illness. Of these:

- 62% were work-related musculoskeletal disorders (WMSDs);
- 25% were stress, depression, or anxiety; and

- 13% were other work-related illnesses.

In 2018, there were 51 000 cases of WMSDs (new or long standing) in the construction industry, with construction workers almost twice as likely to experience a WMSD compared to workers across all industries. The Health and Safety Executive also estimates that around 2.4 million working days (full-day equivalent) were lost each year between 2015/16 and 2017/18 due to workplace injury or work-related illness in the construction industry of Great Britain. Two million of these working days were lost as a result of work-related illness, while 0.4 million were lost as a result of workplace injury, indicating the magnitude of the impact of work-related poor health relative to injury in the construction industry (Health and Safety Executive 2018a).

1.3 The Evolution of Workplace Safety

Over the years, the emphasis on how workplace safety should be tackled has changed, as understanding of the contributory factors to work-related injury has evolved. Some writers suggest the approach taken to managing workers' safety has progressed through a number of discernible periods or ages.

Hale and Hovden (1998) summarize these ages as follows.

- *The 'technical' age*: spanning the nineteenth century until after the Second World War. In the technical age the focus was on technical measures for guarding machinery, stopping explosions, and preventing structures from collapsing.
- *The 'human factors' age*: spanning the 1960s and 1970s. The 'human factors' age considered the main source of accidents to be human error arising from interactions between human and technical factors. The merging of two fields that influenced safety – probabilistic risk analysis and ergonomics – saw the focus shift to human error and human recovery or prevention.
- *The 'safety culture' age*: from the 1980s onwards. The safety culture age developed as it became apparent that matching individuals to technology did not resolve all safety problems. The 1990s saw a growing emphasis on cultural determinants of safety. The main focus of safety development and research shifted to organizational and social factors.

The 'technical age' of safety responded to fundamental changes in agricultural, industrial, and manufacturing processes that began with the industrial revolution. The technical age was principally focused on the development of engineering and technological solutions to newly emerging workplace hazards. However, as shortcomings of focusing heavily on technology were identified, attention shifted to the interface between people and technology. Thus, the second age of safety, as Hale and Hovden designate it, is the 'human factors' age. Within this age, growing prominence was enjoyed by the discipline of ergonomics, a science that deals with designing and arranging things so people can use them easily and safely. Yet again, towards the end of the twentieth century, the focus changed as research revealed the critical role of management and organizational factors in shaping workplace health and safety outcomes. Thus, in the third age of safety, the 'safety culture' age, greater emphasis was placed on organizational and social factors.

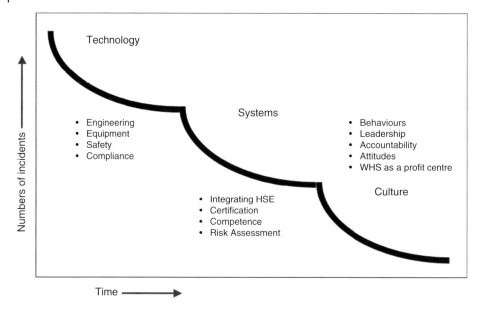

Figure 1.1 The progressive 'ages' of safety. *Source:* Hudson (2007).

Hudson (2007) proposed three slightly different stages in the evolution of safety thinking (see Figure 1.1). These were based on his observation that the focus of safety improvement efforts changes over time in large multinational organizations:

- first, there is an emphasis on technology and the opportunities it affords to reduce injuries and ill-health;
- second, there is an emphasis on implementing safety management systems; and
- third, organizations begin to place greater emphasis on cultural aspects of safety.

Hudson argued that technology and systems-based approaches to managing workplace safety produced significant reductions in incidents (and injuries), but these improvements eventually plateaued.

Thus, the focus on cultural aspects of workplace safety emerged from recognizing that the people within the organization were the missing component in workplace safety processes. A greater emphasis was placed on people and culture in an effort to engage organizational members' 'hearts and minds' in the workplace safety effort, whatever their role or level.

'Step change' models, such as those proposed by Hale and Hovden (1998) and Hudson (2007), reflect a relative change in emphasis in the way that workplace safety has been thought about and tackled over time. These models are focused more on safety rather than health. However, their inherent limitations also have relevance to the management of occupational health. Indeed, these models should not be interpreted too literally because they suggest the benefits to be gained from earlier approaches are already exhausted when a change of emphasis is made. Hopkins (2006b) argues that implicit in these models is the suggestion that the only opportunity to produce further improvement in workplace safety is to stop focusing attention on technologies and systems, and to focus exclusively on culture and behaviour (Hopkins 2006b). Given that technological means for controlling workplace safety (and also occupational health) risks (for example, elimination, substitution, and engineering) are preferable to behavioural means (for example, relying on administrative measures and using personal protective equipment), some of the most effective solutions to the construction industry's WHS problems may lie in better deployment of advanced technologies. Arguably, the best way to do this is to ensure systems and cultures enable technological enhancements and improvements for WHS risk control. There is also an increasing emphasis placed on reducing WHS hazards at source in the design stage of construction project work. Rollenhagen (2010) argued that placing too strong an emphasis on safety culture could potentially discourage technologists from designing better equipment, construction processes, or ways of working. Rollenhagen (2010) also identified the need to improve design organizations' management practices and cultures in relation to developing innovative ways of improving WHS.

In the construction industry, as in other sectors, the analysis of workplace injuries and deaths shows that the underlying causes of incidents include issues of equipment and work process design, the organization of work, and multiple layers of management decision making (Gibb et al. 2014). Even when workers fail to follow work procedures, procedural violations can often be traced back to factors in the organizational and physical work environment (Lingard et al. 2016). This is illustrated in Case Example 1.1.

Case Example 1.1 Multiple Factors in a Work-Related Death

John was in the final year of completing his apprenticeship as a plumber and gas fitter. This meant he was unlicensed and was required to be under a qualified plumber's supervision. The day prior to the incident, John's supervisor requested he (John) attend a caravan park to fit a new gas water heater in a mobile home permanently housed there. Later in the day, after works had commenced, John realized he did not have the equipment he required to complete the job. He returned the following day and began work in a hole that had been dug the previous day, connecting the new gas line to the town mains gas line. It was while undertaking this work that John damaged the mains gas supply and was overcome by gas. Efforts to revive him failed.

During the course of the investigation it was identified that the mobile home's owner had not advised the caravan park proprietor of any works. Further, the proprietor was unaware of the presence of any tradesman onsite, despite having security/access restrictions at the park entrance and a 'sign-in' process in place. This incident reveals a complex interaction between technological, managerial, and behavioural causes. A coronial inquiry found John was not using the correct equipment to allow for safe connection to the gas mains. He had not been trained properly in the work he was asked to undertake, and consequently he failed to recognize what equipment was required to carry out the work safely. At the site, a hole had been dug around the mains pipe and to access this pipe John had to lie on his stomach on the ground and place his head, arms, and the top part of his body into the hole. John was not wearing any protective equipment and did not use the gas detection meter he had been given. Despite being an apprentice, John was not being supervised at the time of the incident and he did not possess sufficient skill or knowledge to carry out this work. The caravan park proprietor did not control access to the site and was unaware that the work was being undertaken. There was no documentation of the work and the owner of the pipeline was not identified or contacted by John or his supervisor.

(*Source:* adapted from Cooke and Lingard 2011)

In designing safe (and healthy) systems of work, it is important to understand and address the interactions between people, equipment, structural components of build-ings and other aspects of the built environment, including underground services, and the processes of construction. A systematic approach to managing WHS is critical to ensuring things are not left to chance and all hazards are identified, analysed, and properly addressed. Glendon et al. (2006) argue the challenge lies in better understanding how technology, systems, and culture can be simultaneously considered, thereby creating the possibility of a more integrated approach to improving workers' health and safety. Glendon et al. (2006) refer to this as 'the integration age' of WHS.

1.4 An Integrated Approach to WHS in Construction

The construction industry is a particularly difficult environment in which to apply an integrated, interdisciplinary approach to WHS. Reasons for this are:

- the industry's fragmented supply arrangements, including high levels of specializa-tion and division of labour;
- use of flexible labour processes that increasingly rely on precarious forms of employ-ment (for example, subcontracting and labour hire); and
- cultural characteristics of the industry that are often driven by client demands and militate against WHS improvements.

Each of these reasons is explored below.

1.4.1 Fragmented Supply Arrangements

Construction industry supply arrangements are highly differentiated and sometimes fragmented, making an integrated approach to managing WHS difficult to achieve. Construction project teams have been described as 'temporary, multidisciplinary and network-based organizations' (den Otter and Emmitt 2008, p. 122) in which multiple specialists work together in a 'web' of interorganizational relationships (Pietroforte 1995, 1997; Nicolini et al. 2001).

Project organizations are vertically segregated as people involved in the initiation, design, production, use, and maintenance of facilities are engaged under separate con-tracts. Depending on the particular project procurement or project delivery model selected, these groups may have limited opportunity to communicate or engage in joint problem solving (Atkinson and Westall 2010). This is a problem because deci-sions made in the early project phases of planning and design are known to have a significant impact on construction workers' health and safety (Hare et al. 2006). Integrating WHS into project planning, procurement, and design activities is discussed further in Chapters 2 and 3.

The traditional separation between design and construction functions in delivering construction projects can impede the development of shared project goals (Baiden and Price 2011) and can negatively impact on project outcomes (Love et al. 1998). Traditional procurement methods militate against the proper consideration of construction WHS issues during the pre-construction planning and design stages, as critical knowledge about construction processes (and their WHS implications) is often not available to decision makers in these early project stages (Yates and Battersby 2003). This is supported by a review of WHS in the UK construction industry that identified the separation of, and poor communication between, design and construction functions as causal factors in construction fatalities (Donaghy 2009).

The acknowledged problems inherent in vertical segregation between contributors engaged to deliver construction projects have contributed to growth in collaborative or integrated forms of project delivery. Integrated project delivery is defined as 'a project delivery approach that integrates people, systems, business structures and practices into a process that collaboratively harnesses the talents and insights of all participants to optimize project results, increase value to the owner, reduce waste, and maximize efficiency through all phases of design, fabrication, and construction' (American Institute of Architects 2007). Such integrated project delivery methods are believed to improve buildability and, by implication, also have the potential to improve WHS (Bresnen and Marshall 2000; Kent and Becerik-Gerber 2010). Technologies such as building information modelling (BIM) have also enabled construction project delivery to become more integrated as information is collected and easily shared between project contributors and across project lifecycle stages (Azhar 2011; Succar 2009).

However, integrated project delivery does not guarantee WHS success (Ankrah et al. 2009). Instead, actual WHS improvements are likely to occur as a result of increased communication and information exchange among project participants afforded by the integrated delivery method. An Australian analysis of the impact of commercial frameworks on construction project WHS performance affirmed this, with one senior industry figure interviewed explaining: 'You can make a huge impact on safety no matter what the commercial framework is. Because it's people generally who are the solution to how we get better at things.'

This quote illustrates the important role played by people and their relationships in driving WHS performance. The distributed nature of project teams can create challenges for a coordinated approach to managing WHS. Distributed teams are those in which some individuals may be co-located, but others are clustered in other locations, preventing regular or routine face-to-face interaction (Stagl et al. 2007). Participants in these distributed teams can make or influence decisions with the potential to impact WHS – sometimes with little or no knowledge of these impacts. In distributed teams there are fewer opportunities to monitor team members' behaviour and provide feedback. Fewer opportunities to observe non-verbal cues can also create ambiguity and reduced situation awareness in members (Fiore et al. 2003). Added to this, construction work is inherently stressful because it is often undertaken under conditions of time pressure, with severe financial penalties for time overruns (Leung et al. 2008; Bowen et al. 2013a,b). Stress can cause people to lose the team perspective and become narrowly focused on the performance of their own individual tasks (Driskell et al. 1999). Further, while working in the temporary construction project environment, participants must also balance the interests of the project with their own individual professional or business interests. All these factors make it difficult to achieve a common purpose and an integrated approach to managing WHS risk.

Establishing shared mental models has been examined as a means of enabling improved team coordination and performance (Salas et al. 2005; Banks and Millward 2007). Mental models are defined as 'mechanisms whereby humans are able to generate descriptions of system purpose and form, explanations of system functioning and observed system states and predictions of future system states' (Rouse and Morris 1986, p. 351). Shared mental models within a team refer to an organized understanding or mental representation of knowledge shared by team members (Cannon-Bowers et al. 1993). In teams with strong shared mental models, members implicitly coordinate their efforts to focus on achieving team goals (Fisher et al. 2012), and teams are most effective when members are able to anticipate and predict other members' needs, identify changes in the task or team, and adjust their strategy as needed. The existence of shared mental models in work teams has been linked to safety performance (see, for example, Smith-Jentsch et al. 2005), but differences between managers and workers' WHS mental models have been observed (Prussia et al. 2003). A study by Lingard et al. (2015d) also revealed that construction project participants (including architects, engineers, and construction and WHS managers) had significantly different WHS mental models. These differences are attributed to variation in experience, education, and professional focus.

1.4.2 Flexible Labour Processes and Precarious Employment

Flexible labour hire practices benefit construction contractors in helping them to cope with changing market conditions and a competitive tendering environment. These practices (including multiple levels of subcontracting and, increasingly, the use of labour hire) have been linked to reduced levels of WHS performance (Mayhew and Quinlan 1997). Quinlan (2011) suggests WHS problems arise in supply and production networks as a result of three factors.

1. Economic and reward pressures that become successively greater towards the bottom of supply chains.
2. Disorganization due to the engagement of many different (often small) businesses.
3. Workers, whose employment is often precarious, working within complex and fragmented production arrangements.

Subcontractors are engaged by principal contractors to undertake a substantial proportion of construction work. In some sectors, principal contractors effectively take on the role of managing contractor and subcontract out all physical construction activity. Subcontractors are positioned at the lower end of the hierarchical structure of contracting and have the highest exposure to hazards and risks (Lingard and Holmes 2001). The low profit margins that result from a competitive tendering system mean subcontractors may be reluctant to invest in WHS.

Many subcontracted workers do not believe legislative requirements adequately address their particular safety concerns, including manual handling injuries and repetitive movement injuries (Wadick 2010). Further, subcontracting often operates on a payment-by-results basis; that is, payment is based on the amount of work completed rather than the time spent on work (Mayhew and Quinlan 1997; Wadick 2010). This arrangement can drive subcontractors to work excessively long hours and take WHS 'shortcuts'. Depending on their employment arrangements, some subcontracted workers in construction may have limited compensation, holiday, sick leave, or superannuation entitlements (Mayhew and Quinlan 1997; Mayhew et al. 1997). Difficult access to compensation, and financial pressures, may cause them to continue working after injury instead of seeking medical treatment. Thus, chronic injuries are common among subcontracted workers, and research indicates many workers take early retirement due to disability caused by injury sustained at work (Mayhew and Quinlan 1997).

Communication between subcontractors engaged to work at a construction site can sometimes be poor and it is the job of a principal contractor to ensure work is properly coordinated so that the activities of one subcontractor do not increase dangers to others. The fragmentation of trade-based subcontractors can also create ambiguity about the boundaries of WHS responsibility (Mayhew and Quinlan 1997). Wadick (2010) reports that subcontractors in the construction industry's residential sector perceive WHS management systems imposed by principal contractors as being heavily paper-based, irrelevant, costly, and ineffective. In some cases, subcontractors distrust these systems, believing them to be driven by the principal contractors' desire to protect themselves from possible criticism or legal liability, rather than a genuine interest in protecting workers' WHS (Wadick 2010).

1.4.3 Cultural Characteristics of the Construction Industry

Christensen and Gordon (1999) explain cultural differences between industry sectors in terms of broader industry imperatives. Gordon (1991) argues that organizational culture is deeply influenced by the characteristics of the industry in which the company operates. Companies in the same industry usually share some common cultural values and practices that are essential for survival in the industry. This is because industry-driven assumptions create industry-wide value systems, which lead companies to develop strategies, structures, and processes consistent with – and not 'antagonistic' towards – the prevailing industry culture.

The construction industry is well known as a male-dominated industry with a strongly masculine culture (Gale and Cartwright 1995; Loosemore and Galea 2008). Mearns and Yule (2009) report that industries characterized by a male-dominated, 'macho', 'can do' culture tend to attract, accept, and retain workers who are inclined to take greater risks. The construction industry follows traditional work patterns and is characterized by a culture of long hours and weekend work, especially for site-based workers. Lingard and Francis (2004) report that, on average, site-based employees in direct construction activity work 63 hours a week, employees in site offices work 56 hours, and employees in the head offices of construction companies work 49 hours. In addition, the project-based nature of construction work, and the uncertainty associated with competitive tendering systems, lead to many workers experiencing a lack of job security, or suffering from frequent relocation as a means of ensuring continuity of employment (Lingard and Francis 2004).

This demanding work environment impacts construction workers' WHS and non-work life in a negative way. Lingard and Francis (2004) found that project-based construction workers experience high levels of work–family conflict and emotional exhaustion as a result of excessive job demands, including long and irregular work hours. In another study, Lingard et al. (2010a) reported Australian construction employees showed higher mean scores for time-based, strain-based, and behaviour-based work-interference with family (WIF) compared with scores reported in other international studies. They found those who work onsite in direct construction activity had higher levels of time-based and strain-based WIF than salaried workers who work predominantly in office-based roles. Long work hours and high work pressure interfere with construction workers' ability to fulfil family responsibilities, and have a detrimental effect on their health and wellbeing.

Dainty and Lingard (2006) report that the need to comply with male-oriented work practices, such as the expectation that workers will work long hours and work in disparate geographical locations, is an impediment to women's career advancement in the construction industry. The under-representation of women in the construction industry means their behaviour is subject to even greater 'time scrutiny' than their male counterparts, increasing the pressures upon women to be available for work at all times. Indeed, in an industry culture that 'glorifies' workers who work as though they have no personal life, it is extremely difficult for workers (male or female) with primary responsibility for caring for children or other family members to manage the demands on their time.

Social and cultural aspects of work are also reported to impact negatively the health and work ability of male, manual/non-managerial workers. Kolmet et al. (2006) interviewed Australian male, manual/non-managerial workers and found a tension between cultural constructs of masculinity (for example, the need to feel 'in control') and low levels of control they have in their work situations. Low levels of job security associated with project-based work and precarious employment arrangements created a sense of disempowerment and resignation to the likelihood of diminished life expectancy. Du Plessis et al. (2013) also describe how 'hyper-masculine' subcultures develop in male, manual/non-managerial work environments. In these subcultures, unhealthy lifestyle behaviours are often inadvertently promoted and workers who seek help with health problems are regarded as 'weak' (Iacuone 2005).

Despite structural and cultural challenges associated with achieving an integrated approach to managing WHS in the construction industry, the potential improvements that could be made by doing so are substantial. Integration is the central theme of this book.

1.5   Structure of the Book

This book describes research undertaken over a 10 year period in the Centre for Construction Work Health and Safety Research at RMIT University. At the time the work commenced, in 2005, Australia was implementing legislative changes that allocated responsibility for construction workers' health and safety to designers. There was growing industry and academic interest in defining 'best practice' in terms of WHS and in exploring the role played by organizational, project, and workgroup culture in shaping construction industry WHS. Each chapter in this book incorporates data collected in collaboration with construction industry partners.

The following chapters examine the underlying need to address, in a more integrated way, the construction industry's relatively poor health and safety performance.

Chapter 2 describes the role clients can play in establishing clear objectives for WHS from the commencement of a construction project, and how they can drive WHS performance through their procurement and project management activities. A Model Client Framework is presented. This framework establishes actions for clients across the project lifecycle that can create conditions within which WHS is integrated into project decision making and management. Evidence is presented relating to clients' opportunities to influence WHS when establishing the commercial arrangements developed to deliver construction projects.

Chapter 3 provides a review of the operation and effectiveness with which WHS is integrated into the design of the construction industry's products and processes. International evidence linking design decision making and WHS outcomes is presented, and policy and legislative responses are explained. Research data is presented to show that early consideration of WHS in project decision making is linked to implementing effective WHS risk-control outcomes. The importance of good communication between project stakeholders engaged in complicated project delivery networks is also discussed in relation to achieving good WHS outcomes. An example is provided of an effective infographic tool for communicating WHS information about construction work processes to design consultants, illustrating the benefits of visual communication in improving the integration of WHS into design decision making.

Chapter 4 discusses the neglected issue of construction workers' health. An integrated model of workers' health is presented. The model links work environment characteristics with personal factors, and links quality of work–family interaction with health outcomes and WHS performance. The chapter considers the impacts of organizational issues, and the quality of jobs and work on the health of the construction workforce. The need to understand construction workers' health in the social ecological context in which it occurs is explained. Research evidence is presented demonstrating that health promotion programmes are likely to produce limited, unsustainable improvements if they target workers' behaviours without addressing occupational, organizational, and environmental factors that contribute to poor health.

Chapter 5 provides a comprehensive review of organizational and project cultures and their potential to influence construction workers' WHS. Particular cultural impediments to WHS are identified, as well as factors that enable WHS. Nine components of culture linked to WHS are identified, all of which can operate at either organizational, project, or even workgroup levels. A maturity model, developed from research conducted in the Australian construction industry, is presented. This model contains descriptive characteristics of a construction organization or project as it progresses through five distinct cultural maturity levels. The model can assist construction organizations to develop organizational and project cultures that enable (rather than impede) WHS.

Chapter 6 discusses using measurement and metrics for WHS. Different types of performance indicators are considered and critically reviewed. Analysis is presented of a five-year dataset, collected at a large infrastructure construction project, showing the relationship between leading and lagging WHS performance indicators. This analysis shows a cycle of reciprocal relationships, calling into question the usefulness and interpretation of some commonly used, so-called leading indicators of WHS performance. Safety climate assessment tools are also discussed as leading indicators of WHS, and longitudinal data collected at five construction projects is presented to illustrate the value of assessing climate changes over time in the evolving construction project environment. The chapter considers the need to develop appropriate metrics for measuring the effectiveness of upstream WHS activities, including integrating WHS into design decision making. A practical tool is presented, based on the hierarchy of control, that can measure the effectiveness and impact of design decision making on WHS.

Chapter 7 presents a discussion of the role of rules and engagement in ensuring people work in healthy and safe ways. The chapter considers assumptions about human behaviour that underpin the operation of WHS management approaches, and two contrasting perspectives are presented about human error and how to achieve good WHS. Arguments are presented that position human error as a symptom of something that is wrong in a system of work, rather than a cause of incidents. Research is presented that used participatory video to understand the reasons why construction workers break WHS-related rules in their everyday work, and the findings are used to mount an argument for engaging workers in designing work procedures to ensure rules are practical and make sense in the work environment. A rule management process is presented to ensure rules are well designed and remain relevant to a particular work environment. This process acknowledges procedures and rules cannot to apply to all situations and adaptations are necessary, but these adaptations must also be managed carefully through mechanisms that match workplace cultures and workers' capabilities.

Chapter 8 provides detailed analysis of the problem of work-related musculoskeletal disorders (WMSDs) in the construction industry, and presents a case for a holistic approach to addressing this problem. The chapter draws together themes from the preceding seven chapters to describe how WMSDs are caused by complicated interactions between characteristics of physical work tasks and psychosocial risk factors in the work environment. Ergonomic interventions used to reduce the risk of WMSDs in construction are described, and opportunities are explored for risk reduction through design of construction products and processes. The chapter describes using participatory ergonomics to engage workers in redesigning work processes with the aim of reducing WMSD risk. Examples are presented of this approach applied in the construction industry.

Chapter 9 concludes the book. It discusses the different aspects of integration covered in earlier chapters, and how each aspect can be used to inform the development of strategies to improve construction industry WHS. Emergent trends in WHS practice are critically reviewed and suggestions made about future directions for WHS policy, practice, and research.

Discussion and Review Questions

1 Is too much effort placed on preventing high-frequency/low-consequence safety incidents? Why/why not?

2 To what extent are supply arrangements fragmented in the construction industry? What are the implications for the management of WHS?

3 How could a more integrated approach to WHS be achieved in construction? What would be the potential benefits of such an approach?

2

The Client's Role in Improving Workplace Health and Safety

Helen Lingard, Nick Blismas, Tiendung Le, David Oswald, and James Harley

School of Property, Construction and Project Management, RMIT University, Melbourne, Victoria, Australia

2.1   Can Clients Influence Construction Workers' Health and Safety?

Contemporary models of accident causation recognize the importance of organizational issues and management actions in contributing to workplace accidents (Reason 1990, 2008). The analysis of workplace health and safety incidents in construction projects reveals that incidents can sometimes be, at least in part, attributed to professional or managerial failures arising well before work commences on site (Bomel 2001; Suraji et al. 2001; Health and Safety Executive 2003). Consequently, there is a growing trend in work health and safety (WHS) policy and practice for management responsibility to be driven up the supply chain, to rest with construction clients, owners, and other parties involved in the planning and design of construction projects. (The designer's role and responsibilities in relation to construction workers' health and safety are discussed in detail Chapter 3.)

Client requirements have been identified as a possible causal factor in construction site safety incidents (Health and Safety Executive 2003). It has also been suggested that client involvement in project WHS activities can improve performance in construction projects (see Huang and Hinze 2006a,b; Winkler 2006). The belief in clients' potential to influence construction workers' health and safety has led some countries to establish specific responsibilities for construction clients in WHS legislation. For example, the UK's Construction Design and Management Regulations (The Construction (Design and Management) Regulations 2015 (UK) 2015), establish a client's duty to make suitable arrangements for managing a project, and to maintain and review these arrangements throughout, so the project is carried out in a way that manages WHS risks. For projects involving more than one contractor, the client is also required to appoint a principal designer and a principal contractor, and to make sure these parties carry out their duties properly (Health and Safety Executive 2015). Examples are becoming more prominent of construction industry clients taking a more proactive stance to improve WHS in construction projects, particularly in delivering major public infrastructure projects (Eban 2016).

In the USA, clients do not have a legislative duty but the role of the client is still acknowledged to be important. Thus, the American Society of Civil Engineers' Policy Statement on Construction Site Safety (ASCE 2012) specifies clients' responsibilities as:

- assigning overall project safety responsibility and authority to a specific organization or individual (or specifically retaining that responsibility);
- designating an individual or organization to develop a coordinated project safety plan and monitor safety performance during construction;
- designating responsibility for final approval of shop drawings and details through contract documents; and
- including prior safety performance as a criterion for contractor selection.

These responsibilities reflect an expectation that clients take an overall coordination role, engage competent design consultants and contractors, and ensure suitable arrangements are made for managing WHS. However, a case can be made for more active client involvement in considering WHS in their procurement and project management activities.

In a construction project the client is analogous to senior management within a single organization. The client defines the need for the project, is responsible for specifying project requirements, and dictates constraining factors like the project schedule and budget (Levitt and Samelson 1993). The client sets the 'tone' of a construction project, and articulates the relative importance of major project objectives, such as time, cost, quality, and WHS. Clients make key decisions concerning project objectives, project budget, and performance criteria. They determine project timelines, which can create the type of pressures and constraints known to have a significant impact upon workers' health and safety during construction. Gibb et al. (2014) explain that all these decisions can potentially impact the health and safety of construction workers.

Specific initiatives designed to improve the construction industry's performance have focused attention on worker behaviour, management systems, aspects of organizational culture, and, more recently, design decision making. Thus far, relatively little attention has been paid to the ways in which clients can drive practical improvements in construction workers' health and safety. Therefore, the relationship between the client's involvement and the level of WHS performance in a project is not well understood (Niu et al. 2015).

Based on client WHS practices, Huang and Hinze (2006a,b) attempted to develop predictive models of project safety performance. They adduced preliminary evidence that certain client actions are associated with enhanced safety performance. However, this research was limited in a number of important respects:

(i) information about the range of client WHS activities was restricted;
(ii) the measure of implementation of client WHS activities was a blunt 'binary' measurement which did not reflect the quality of implementation; and
(iii) measurement of client actions was based solely on 'self-reported' data from client organizations.

Spangenberg et al. (2003) have conducted an analysis of the impact of a client-led health and safety programme implemented during construction of the Øresund rail link between Denmark and Sweden. They found that a multifaceted programme produced a 25% reduction in the number of injuries resulting from safety incidents. The programme included a large-scale information campaign, a twice-yearly monetary award, and specific themed campaigns aimed at improving workers' health and safety-related behaviour. However, the authors note that the programme's impact may have been limited because:

- it focused too heavily on trying to change attitudes towards WHS (through providing information), rather than changing health and safety practices; and
- contractors were only involved in the project for relatively short periods of time, limiting their exposure to the campaigns.

Thus, while there is some emerging evidence to suggest construction industry clients can and should do more to drive improvements in workers' health and safety, it is not entirely clear what client activities produce the best WHS outcomes.

This chapter discusses ways in which clients can engage in WHS in the projects they procure. The chapter describes a set of guidelines developed to help Australian Government agencies manage WHS in the projects they procure. These guidelines reflect a lifecycle approach to managing WHS in which clients are actively engaged in project activities from the beginning to the end of a construction project. A case study is presented which documents the implementation of this lifecycle WHS management process. We then discuss the ways that commercial frameworks used to deliver projects impact health and safety performance. We draw on cross-case comparative data to illustrate how WHS performance is driven by clients' selection of project delivery method, specification, and measurement of key performance indicators (KPIs) for WHS, and methods for remunerating consultants and contractors. We also discuss the need to position commercial frameworks in the broader construction industry context, structure, and culture.

2.2 The Role of Governments as Policy Makers and Major Purchasers

In her review of deaths in the UK construction industry, Rita Donaghy argued that 'public procurement is important because of its size and its potential for insisting on driving up standards including health and safety' (Donaghy 2009, p. 12). The *Australian Work Health and Safety Strategy 2012–2022* ('The Strategy') also recognizes the potential role of public sector clients in driving WHS performance improvements. The Strategy specifically calls for using commercial relationships to improve WHS and for Australian governments to use their investment and purchasing power to improve WHS (Safe Work Australia 2012, p. 11).

The Strategy identifies governments as having a range of tools they can use to change WHS behaviours. Governments can influence change through policy development and in the programmes and services they deliver. They are also major purchasers of products and services. Governments are consequently in a powerful position to drive health and safety improvements 'by incorporating work health and safety and safe design requirements into government investment, procurement arrangements and contracts' (p. 10). In doing so, it is expected that government agencies actively encourage suppliers to improve WHS health performance in delivering products and services.

In response to the *Australian Work Health and Safety Strategy 2012–2022*, the Australian Government aims to achieve world-class WHS in the building and construction industry. It is incumbent on Australian Government agencies – as clients – to drive positive WHS performance through their procurement and project management processes. The Model Client Framework (MCF), described below, is one initiative that takes a lifecycle approach to client engagement in WHS.

2.3   The Model Client Framework

In 2007, the Office of the Federal Safety Commissioner commissioned Lingard, Blismas, and others to develop a set of guidelines, known as the Model Client Framework (MCF), to help government agencies embed WHS into construction procurement and project management processes (Lingard et al. 2009a). The resulting framework consists of five booklets which establish principles for managing project WHS and define key management actions (KMAs) for implementation throughout the project planning, design and procurement, construction and completion stages.[1] Tools and resources are provided to support the practical enactment of each of these KMAs. Through implementing the MCF, Australian Government agencies are striving to ensure major stakeholders involved in planning, design, and execution of construction work collaboratively allocate responsibility for WHS and integrate health and safety considerations into all project decision making. The Office of the Federal Safety Commissioner developed the principles underpinning the MCF. The principles are summarized below.

2.3.1   Principle 1: Develop a Project Culture that Enables WHS

Model clients should demonstrate a tangible commitment to WHS within their own organization and across the building and construction industry. The Model Client process is driven by the overarching aim of creating a project culture that is shared by project participants and which enables exemplary WHS. The culture is one that expects all participants to treat WHS as an integral part of managing the project, and that health and safety objectives stand on a footing similar to other project objectives, such as quality, cost, and timeliness.

The project culture should also emphasize collaboration and teamwork between all stakeholders, with the aim of establishing a shared set of values, assumptions, and beliefs that reflect a strong commitment to workers' health and safety. Effective communication, confidence, and trust will be integral parts of a project culture that enables WHS. The client should encourage openness in error and incident reporting so they are regarded as opportunities for learning and improvement.

---

1 www.fsc.gov.au/sites/fsc/resources/az/pages/themodelclient-promotingsafeconstruction

2.3.2 Principle 2: Leadership and Commitment

Model clients should demonstrate leadership in WHS at all stages of a project lifecycle by acting as exemplars in their relationships with other industry participants. Leadership and commitment by a model client can be demonstrated by:

- incorporating WHS considerations at every level of decision making in construction projects, from procurement to completion;
- articulating a WHS vision and ensuring contracts for the supply of goods and services clearly reflect the expectation of high standards for WHS; and
- actively monitoring WHS through all project lifecycle stages, acknowledging good health and safety performance, and correcting substandard performance.

2.3.3 Principle 3: Develop Cooperative Relationships

Model clients should strive to develop cooperative business relationships to ensure time, cost, and quality objectives do not compromise a commitment to workplace health and safety. A model client can demonstrate development of cooperative relationships by:

- facilitating the establishment, at the earliest stage practicable, of an integrated project WHS management team – including designers, contractors, and model client representatives;
- ensuring their managers lead by example and communicate, throughout all stages of the project lifecycle, the importance of WHS in interactions with all project stakeholders, including designers, contractors, and suppliers; and
- establishing long-term relationships with service providers to support the development of WHS capability within the supply chain.

2.3.4 Principle 4: Promote WHS in Planning and Design

Model clients should ensure safe design and constructability are considered at the planning and procurement stages of a project to reduce or eliminate hazards and control risks before construction commences. A model client's planning and design processes ensure WHS issues are considered by:

- clearly specifying WHS as a criterion in the project design brief and selecting design consultants who have a demonstrated capacity to consider WHS risks;
- collaborating with stakeholders to eliminate or reduce WHS risks by making decisions based on careful consideration of the WHS implications of available design options; and
- overseeing WHS design reviews at appropriate stages during the project life – especially where design changes are proposed during the construction phase.

2.3.5 Principle 5: Consult with and Communicate WHS Information to Project Stakeholders

Model clients should ensure effective consultation and communication arrangements are in place so that all stakeholders are aware of WHS considerations and of their responsibilities. A model client makes sure health and safety information is communicated to all project stakeholders by:

- ensuring clients pass on information to designers and designers pass on information to contractors – and further through the supply chain to subcontractors – about WHS risks associated with proposed materials, substances, or construction methods;
- ensuring mechanisms are established to convey and record WHS risks to all stakeholders throughout the project lifecycle; and
- facilitating bottom-up communication of safety issues and consultative processes to enable worker participation in making decisions that impact upon WHS.

2.3.6   Principle 6: Manage WHS Risks and Hazards

Model clients should ensure a systematic approach is taken to managing WHS risks and hazards. A model client manages health and safety risks by ensuring that:

- hazards are identified at all stages in the project lifecycle and health and safety risks are systematically assessed and controlled;
- identified risks are eliminated or, where elimination is not practicable, reduced so far as possible – preferably through implementing technological controls; and
- project decision making that could impact upon WHS risk involves input from those people or groups of people who could be affected by that risk.

2.3.7   Principle 7: Maintain Effective WHS Measures Across the Project Lifecycle

Model clients should ensure they maintain effective WHS measures across the construction project lifecycle and that they respond to changes in the construction environment. A model client will maintain effective WHS measures and continuously improve WHS performance, by:

- requiring regular reporting of project WHS performance, using both 'leading' and 'lagging' performance measures, and by conducting regular, ongoing, and project completion reviews;
- using health and safety performance data to identify problems and implement improvement strategies before incidents occur; and
- seeking feedback from service providers and contractors on their own performance as a client and acting on identified weaknesses.

2.3.8   Principle 8: Monitor and Evaluate WHS Performance

To compare and improve health and safety performance, model clients should monitor, report, and benchmark WHS at the site, project, and company levels. A model client will monitor and evaluate health and safety performance by:

- establishing meaningful and reliable performance indicators to measure project health and safety performance against industry benchmarks;

- providing stakeholders with accurate comparative health and safety performance information; and
- ensuring WHS performance measures cover the whole project lifecycle and include feedback from workers and subcontractors.

Traditionally, construction clients have adopted a 'hands off' approach to managing WHS during a project's construction phase. However, there is emerging empirical evidence to suggest more active engagement and client involvement in construction project WHS activities can improve performance. Smallwood et al. (2009) argued that WHS performance improvements depend on the extent to which clients provide leadership on WHS matters. Zhang et al. (2015) identify clients' WHS leadership behaviour as an important factor in driving positive and supportive climates for WHS in construction projects. Votano and Sunindijo (2014) report that project-level safety climates are more positive when clients:

- record risk information;
- conduct design safety reviews;
- include safety in contract documents;
- set project safety targets;
- participate in site-based safety programmes;
- review and analyse safety data;
- appoint a safety team;
- select safe designers;
- select safe contractors;
- specify in tenders how safety is to be addressed; and
- perform regular checks on plant and equipment.

Wu et al. (2016) also report that a client's leadership has direct effects on contractors' WHS approach and effectiveness. Clients, therefore, can usefully influence WHS by engaging in joint problem solving and acting as approachable mentors to construction teams and contractors (Wu et al. 2015). Being approachable, participating, supporting, and collaborating with contractors can help develop cooperative relationships and facilitate improved WHS performance. Thus, an active client is:

- embedded in the project;
- involved in the contractors' WHS programmes and initiatives;
- engaged with the contractor to problem solve;
- offers help and support;
- provides feedback, shares knowledge;
- helps creates common goals; and
- allows opportunity for innovation.

The notion of being a WHS-active client was raised in an interview with the Safety Director of a government agency delivering a large rail construction project.

> It means that we're involved and engaged during delivery. We're not taking the thin approach which is all about risk transfer and stepping back and watching. We remain active and engaged during delivery. We do things that contribute to

better safety outcomes… We seek to balance our need to meet our legislator obligations against a need to inspire and influence the industry to perform better in the safety space than they have previously. And we recognise that in order to do that you've got to be an active and engaged client during delivery.

Zhou et al. (2012) observed the importance of multiparty collaboration for safe construction, as the client, designer, principal contractors, and subcontractors all have important roles to play in providing safe and healthy workplaces . For instance, as initiators of projects and in their position at the apex of the contractual hierarchy, clients have an important role to play in promoting WHS in planning and design, and in ensuring the design brief emphasizes and requires that WHS hazards are eliminated or risks reduced to the extent possible when design decisions are being made. Sperling et al. (2008) also identify the important role clients can play in focusing the efforts of design consultants and contractors on WHS.

The growing volume of evidence of a client's opportunity to influence WHS is entirely consistent with the Model Client Principles. Table 2.1 further highlights how the Model Client Principles are currently put into practice by clients in the procurement and project management of large infrastructure projects in Australia. The examples were gleaned from interviews with representatives of client and construction organizations in Victoria and New South Wales.

Table 2.1 Model Client principles in use.

Model Client principle	Description of implementation
Principle 1: Develop a project culture that enables WHS	• *Project director:* 'As a client, they were interested in us instigating a safety culture program… but they didn't actually drive it themselves; they were interested in us undertaking it and taking charge of it'. • *WHS manager:* 'They're [client] not all about compliance. They're actually about having our leaders out visible and engaging out onsite with the workforce, so they're actually understanding the culture that drives the performance'.
Principle 2: Leadership and commitment	• *WHS manager:* '… so leadership on the site that undertakes regular visits and connects with the workface. So that would involve clients or our project leadership team or our senior leaders and that is purely understanding how the job is performing and talking to people onsite'. • *Contractor WHS general manager:* 'Where we've seen [WHS] work really well is where our clients have got on board and they're actually participating in our programs'.
Principle 3: Develop cooperative relationships	• *WHS director:* 'The environments are too complex. When you're putting, like we just done recently, package one, 1,000 people going through the site in a 24 hour period, they had something like 1,700 pieces of plant go through the job in that 37 days. You've [client and contractor] got to work together'. • *WHS director:* 'We ran a safety subcommittee… that was a monthly meeting where the safety manager from each package of work came in and we had a roundtable discussion about what was happening on each package. We shared incident information, trend information, and started to share initiatives. So if one package was running a sun smart initiative, we wouldn't say to the other five packages, "Run your own initiative,"

Table 2.1 (Continued)

Model Client principle	Description of implementation
	we'd share and use that initiative. Contractors were willing to share what they were doing and whether it worked or not, what the issues were with other contractors. We essentially became the keeper of those knowledge management papers, the keeper of those initiatives, the keeper of the information that was being shared.'
Principle 4: Promote WHS in planning and design	• *Project manager:* 'The client has carriage of any project earlier than anybody else. Generally speaking, a client comes up with a reference design on which a tender price would be based. So it's absolutely incumbent on a client when they come up with that reference design is to come up with something which is buildable. And buildable in a safe way.'
Principle 5: Consult with and communicate WHS information to project stakeholders	• *WHS director:* 'As a client, we're in a position to identify issues occurring within projects on that program, and we can suck those issues out of the projects and share them across the program. So there's an umbrella or a helicopter role around sharing information and connecting people, connecting issues, connecting solutions. So very important.' • *WHS director:* 'You can't just simply say, "We want the best performance," or "We want exemplary or best practice performance," without engaging and working with the contractors to help make that happen.'
Principle 6: Manage WHS risks and hazards	• *Project manager:* 'If you [client] looked at some of the high risk activities, so plant–people interface or those sort of activities. You could then say, "Well, what is best practice around the world? What is some of the new technology that's coming into play?" And start implementing and start looking at ways that you can, I mean, there were projects where they [client] basically said, "Every bit of gear that comes on this site has to have XYZ, don't care, you know, if you want to work with us, you've got to have that gear." It's those sort of positive, proactive actions that you take.' • *Safety director:* 'We basically said, "Yes, this contractor is low bid, so on paper it's offering value for money for the State." [But] we saw the contractor was high risk. And it was early in the job so we thought, "Okay, we can buy risk by taking the cheap price. Let's manage risk by taking the next price," which was a contractor that offered more of a holistic offering with [WHS] systems, supervision, better on paper injury performance, all that sort of stuff.'
Principle 7: Maintain effective WHS measures across the project lifecycle	• *Project director:* '… but it was good to actually wrap it into a measure that actually people understood, "Okay well, we're going well/we're not going well." You know, "We've had a dip in performance. What does that mean? How are we going to step it up again this month?"'
Principle 8: Monitor and evaluate WHS performance	• *WHS manager:* '… we had to hit targets. We were audited every three months. So, every quarter we were audited to make sure that we'd hit. And the client would come out and say, "Righto, where are you at?"' • *Client contract manager:* 'So for around safety we had a KRA [key result area] called Our People and Our Workplace, with a minimum condition of satisfaction was that no one gets harmed as a consequence of any of the project activities, and that constructive cultures are the basis of our alliance. That was just a minimum condition of satisfaction. Then you go down to KPIs [key performance indicators]. And we had a KPI called Constructive Safety Culture, so that was a measure of the organisational cultural inventory at the beginning of the project and the subsequent shift towards constructive inventory at the end of the project through behavioural change…'

A significant international project that exemplified strong and active client leadership was the London 2012 Olympic construction programme. The WHS arrangements implemented at the London 2012 programme of works are described in Case Study 2.1.

Case Study 2.1 Delivery Partner and NEC3 Forms of Contract at the London 2012 Olympics

The London 2012 Olympic and Paralympic Games construction programme included the Park, the largest urban regeneration project in Europe; the Village, Europe's largest new housing project; and several other sites remote from the Park. The Olympic Delivery Authority (ODA) was established to ensure the venues and infrastructure needed for the Games were delivered on time, to budget, and fit for purpose. From the outset, the ODA established six priorities against which successful delivery of the works would be measured. 'Health, safety and security' was one of the priorities. The London 2012 construction programme was 'the first publicly funded construction programme to publicly commit to no fatalities.

As a representative of the ODA's WHS Management team explained: 'The whole point about the ODA set up was to be fairly light on its feet and to look at mechanisms for leveraging health and safety performance rather than actively driving it itself.'

Subsequently, the ODA appointed a delivery partner to take charge of the work to deliver the project and manage the supply chain, while the ODA concentrated on managing relations and stakeholder satisfaction to drive delivery. To create a mutually successful partnership, the ODA Delivery Partner structure ensured that the success of the ODA, and achievement of its objectives, were aligned directly to the delivery partner's financial and reputational success. Furthermore, the benefits of establishing a long-term relationship, and the opportunity to improve practices and outcomes across a range of packages, provided the incentive for both parties to work together to provide better value for money (Jacobson 2011).

The ODA developed the 'Health, Safety and Environment (HS&E) Standard' which from the outset clearly communicated the client's requirements and objectives to those delivering the project. The Standard outlined HS&E expectations and requirements for all staff, stakeholders, and suppliers. It applied to all design, engineering, construction, and maintenance works commissioned by the ODA. Apart from requiring contractors and suppliers to comply with HS&E legislation, the Standard also encouraged them to seek out and apply industry best practice to their works. As a representative of ODA's Health and Safety (H&S) management team explained:

> … so right from the beginning the leadership commitment to a high performance in health and safety was woven into the way in which we procured the supply chain. And I think that in a lot of cases what people are doing now is weaving health and safety in with extensive documentation, huge numbers of questions… We were doing it the other way round. We were declaring what we were committed to and asking the contractors who were bidding, "What will you contribute to enable us to do that?"

The ODA and its delivery partner played a key part in developing a positive enabling culture for workplace health and safety at the Olympic Park. Recognizing their influence on the supply chain in terms of setting out programme priorities, the ODA and its delivery

Case Study 2.1 (Continued)

partner required all Tier One contractors (that is, primary contractors with overall responsibility for individual projects) to subscribe to the HS&E Standard and regularly report to the ODA Board on their HS&E performance. Each Tier One contractor was required to:

- have a behavioural safety management system in place;
- adopt a 'no blame' culture;
- have effective communication arrangements to inform all site personnel of key issues; and
- consider introducing reward and recognition programmes to incentivise workers to contribute to good health and safety.

The ODA also focused on working with leaders through the supply chain and engaging them on shared objectives while empowering them to develop their own good practice and drive their own performance. This allowed the contractors to use and develop their own company processes while committing to the client's objectives: '… the argument was that we were going to the marketplace to try and find the best and we wanted the best to bring what they had to offer to what we were doing'.

WHS was considered an essential driver of efficiency and performance. High performance was expected to be achieved through partnership, respect, trust, and open communication:

> … so we were arguing that health and safety was an essential driver of efficiency and performance but we did turn it the other way round. Because the incentives, pain, gain, sharing et cetera, were associated with delivery on time, to quality and within budget, and there were incentives associated with that, we did put in penalties which said you would share less of that incentive if you had sacrificed health and safety on a temporary basis in order to achieve that high performance. But I can't remember situations where those penalties were ever activated because the performance, what we discovered in practice was what we honestly believed intrinsically and upfront… which is that if you are running a program really effectively you can't tease out health and safety.

Client representatives were embedded within the project teams. Thus, expectations for WHS during construction were built into contracts across the supply chain. The leadership team also involved senior representatives of the suppliers directly contracted by the ODA. As a representative of the WHS Management team explained: 'The leadership within the ODA, the delivery partner and then the individual principal contractors, and then their supply chain, was actually key to kind of liberating the [WHS] commitments'.

To drive up consistency and quality in delivery, the New Engineering Contract Version 3 (NEC3) was adopted. NEC3 was considered appropriate as it supported both the partnering approach and the collaboration that the ODA was seeking. A representative of the H&S Management team comments on the effect of the commercial framework on driving WHS activities/behaviours:

> The commercial framework was absolutely fundamental but it wasn't the visible driver of health and safety; it created the context. It gave us the room within which

(Continued)

Case Study 2.1 (Continued)

we could forge those partnerships and maintain that conversation through the works… it ended up being collective rather than kind of client driven and I think the commercial framework we'd adopted made it easier to do that because the NEC3 form of contract encourages that open discussion about change, rather than the client imposition and the willingness to accept the hit of the variation claim.

The selection of the contractual framework was believed to be instrumental in creating favourable conditions for achieving high performance. This was achieved by avoiding disputes, providing a fair basis for compensation and rewards, and clarifying the priorities and expectations through the supply chain. Commenting on the role of contractual framework in driving WHS performance, a representative of the management team explains:

So the commercial frame that we worked with – you know NEC3 which is based upon a very clear approach to early warnings and dealing with compensation events – the way in which you don't allow these things to fester but you have a program that is much more based upon open communications and honesty. And the way in which you treat people is reflected in how you expect them to then perform. I think that was in the DNA of the ODA right from the get go, and I think that really mattered. So it wasn't that the contractual framework was expected to automatically act as a magic wand to deliver high performance. It was that the contractual frameworks were selected and executed very consciously in order to achieve high performance.

The ODA mandated the use of a Safety Climate Tool (SCT) across companies working on the Park. This demonstrated commitment to WHS and made it possible for ODA to gain an insight into the prevailing cultures within the programme of works. The SCT is in the form of a survey that captures workers', supervisors', and managers' perceptions of WHS in relation to eight factors: accidents and near miss reporting; organizational commitment; health and safety oriented behaviours; health and safety trust; usability of procedures; engagement in health and safety; peer group attitude; and resources for health and safety. Contractors were required to complete the SCT at various intervals while working on the Park. This was overseen by the ODA, and resulted in almost 10 000 responses across 20 companies from 2008 to 2011.

In addition, the ODA and the delivery partner required Tier One contractors and designers to self-monitor and submit monthly reports on their efforts to achieve high HS&E standards, and to eliminate accidents, incidents, and significant near misses. Early on, the ODA made efforts through communication campaigns to explain and incentivise the objective reporting of leading and lagging KPIs by the contractors, particularly for near-miss information (Health and Safety Executive 2012).

Eventually, after 62 million hours of work, construction of London 2012 was the first construction programme in the history of the Games completed without a fatality. The onsite accident frequency rate was 0.17 per 100 000 hours, far below the UK building industry average of 0.55 at the time, and less than the average rate of 0.21 for all industries across the UK. There were 22 periods of a million man hours worked without an injury accident reportable under the Reporting of Injuries, Diseases and Dangerous Occurrences Regulations (RIDDOR).

(Additional material sourced from: http://learninglegacy.independent.gov.uk)

2.4   The Model Client Project Process Map

At the heart of the MCF is the Model Client project process map (depicted in Figure 2.1). This process map was based on the generic design and construction process protocol (Kagioglou et al. 1998). The process protocol covers the whole life of the project, from the conception of need to the operation and maintenance of the completed facility. As the authors comment: 'This approach ensures that all the issues are considered from both a business and a technical point of view, as well as ensuring informed decision making at the front-end of the design and construction development process' (Kagioglou et al. 1998).

The Model Client process is divided into four project stages. These are shown across the top of the project process map. They cover the project development stages of planning (Stage A), design and procurement (Stage B), construction (Stage C), and completion (Stage D). The design and procurement stage is separated into two sub-stages: the first covers conceptual design and production design and the second covers procurement.

Each of the stages is further subdivided into phases. The phases are project development steps that occur within each stage of the project. They are shown in the boxes across the top of the page, in the sequence they occur during project development and delivery. For example, in the planning stage, the following phases of activity are undertaken: demonstrating the need; conception of need; outline feasibility; and substantive feasibility and outline approval.

Between each stage (and sub-stage) of the Model Client process map are stage reviews, denoted by vertical bars in Figure 2.1. The purpose of these stage reviews is to ensure all WHS actions have been completed prior to moving to the next project stage. In this way, stage reviews act as 'gateways' in the project process. Before progressing to the next stage of the project, each stakeholder can check to see that all WHS actions from the preceding stage have been completed. Stage reviews also provide an opportunity for project participants to reflect upon the WHS processes and outcomes of the preceding stage and to feed forward important WHS information for use in future project stages.

The project process map provides a common framework for managing and controlling a project, such that the entire team works together to reduce WHS risks to construction site workers, building occupants/users, and maintenance personnel.

The Model Client project process map specifies a number of key management actions (KMAs). These are the actions a model client would be expected to undertake during each stage of a construction project. The position of the KMA in the project process map indicates in which project phase or phases the KMA should occur. For ease of reference, KMAs are numbered sequentially and each KMA is documented using a standard layout. This layout includes:

- 'action', which describes what has to be done;
- 'description', which provides a short narrative of the rationale for the action, covering aspects such as who is responsible, its importance, and some suggested strategies for consideration;
- 'key benefits', which provide the reasons why the action is effective;
- 'desirable outcomes', which describe the behavioural and procedural changes created by implementing the action;
- 'performance measure', which describes the outputs that can be measured and recorded as evidence the action has been implemented successfully; and
- 'documents', which list the suggested documentation that assists in effectively implementing the KMA.

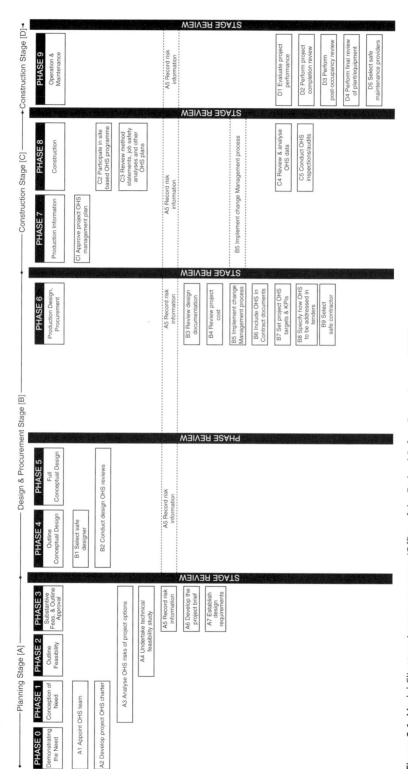

Figure 2.1 Model Client project process map (Office of the Federal Safety Commissioner 2007).

The MCF has been implemented in large-scale infrastructure construction projects. Case Study 2.2 describes the use of the MCF in a large rail construction project.

Case Study 2.2 Practical Implementation of the Model Client Framework

The Office of the Federal Safety Commissioner's MCF has been, and continues to be, applied successfully in Australian multibillion dollar programmes of work for transport infrastructure construction. These programmes are managed by state government authorities established to deliver the programmes, with work predominantly delivered via collaborative alliances or 'design and construct' contracts.

Initially the MCF was used in delivering a programme to construct 47.5 km of rail track through Melbourne's western suburbs, which also included constructing new stations and new platforms at existing stations.

At this programme of work, the client's safety management system was developed following a review of MCF requirements, as well as reviews of:

- project safety risk registers;
- government departmental policies and procedures;
- relevant occupational health and safety and rail safety legislation; and
- relevant requirements of the rail operator safety management system.

According to a Senior Safety Manager involved in delivering the programme of work: 'The MCF formed a sound process structure for the safety management system, and the MCF criteria were interpreted and applied in the program context. The MCF criteria were applied at a high level, without strict application of all process detail or templates provided in the criteria.'

This structure worked well in the first programme of works, and the client safety management system was updated and improved for a second programme of works, commencing in 2016, which involves staged removal of 50 level crossings across Melbourne's rail network.

The Senior Safety Manager engaged in both these programmes of construction work explained:

'The enhancement, not covered in the MCF criteria, was the client management approach. From inception, both the leadership and safety teams set out to inspire exceptional safety performance on the programs that would be recognized throughout the industry for years to come.

'This required adoption of a non-adversarial, collaborative approach, with both design and construct contractors and alliances. This was supported by proactive safety support to internal delivery and contractor safety teams at the respective client delivery authorities. This resulted in the application of ever advancing collaboration and leadership processes at every opportunity. It also demanded a unified commitment from all stakeholders supported by an environment of trust and transparency.

(Continued)

Case Study 2.2 (Continued)

> 'From a legislative perspective, the delivery Authority's involvement on the project did not include management and control of worksites. However, the Authority maintained a commitment to provide the necessary leadership required to inspire and influence contractors and other stakeholders to, not only promote, but to value exceptional safety performance. Industry-leading safety performance was achieved on the first program and has improved again on the current program.'

In both programmes of work, the client organization dedicated considerable resources to their oversight of health and safety. In the early stages of the first programme, safety resources consisted of a Safety Manager and a Safety Advisor. Their activities were primarily centred on basic compliance, including inductions, providing health and safety training, and developing documentation.

However, once the safety management system was finalized, MCF requirements were met, and the Authority's vision for best practice safety was established, the base team of two grew to ten at the peak of the works.

Project complexities, together with delivering the safety management system, required allocating at least one dedicated full or part-time health and safety professional (from the client side) to each work package. This was necessary to ensure the project team and work package contractor had a consistent first point of contact for health and safety. In addition, and dependent on size, complexity, and geographic spread of each work package, a second client employee was sometimes allocated to provide core administrative support to the client health and safety function.

The Model Client process identifies multiple actions clients can take to drive WHS performance in projects they procure. Considering WHS in tendering and contractor-selection decisions is one important point of influence. Different approaches have been taken to linking tendering opportunities with WHS performance. For example, since 2004, Australian legislation has linked WHS performance and practices to tendering opportunities for publicly funded construction projects. Under the Australian Government Building and Construction WHS Accreditation Scheme, head contractors awarded construction work funded directly or indirectly by the Australian Government (above a relatively low threshold value) must be accredited. Accreditation involves submitting an application, followed by on-site auditing of a construction company's WHS management practices and performance. The Scheme focuses strongly on evidence of a systematic approach to WHS, but has been criticized for placing too great an emphasis on WHS-related documentation.

Case Study 2.3 describes a different strategy, implemented in the Australian Capital Territory (ACT). It is designed to ensure only contractors with good WHS records are awarded Government-funded construction work.

Case Study 2.3 Active Certification of Contractors in the Australian Capital Territory

Following a spate of fatalities in the construction industry, the ACT Government launched an inquiry to examine compliance with, and application of, WHS laws in the ACT construction sector. The inquiry's aims were to inform the ACT Government, employers, workers, and the general community about the state of compliance with health and safety laws in the ACT's construction sector and to identify further measures which could improve compliance. The resulting report, *Getting Home Safely* (WorkSafe ACT 2012), identified the potential for ACT Government agencies to drive health and safety improvements in their construction project procurement practices. The report argued that 'as well as a "push" effect, through its role as regulator, Government, through its role as a major client with significant purchasing power, can also have a "pull" effect on the local industry'.

Getting Home Safely identified public procurement as providing an important opportunity to set a high standard for WHS performance in the construction industry. Further, by raising WHS performance standards in public sector projects, it was anticipated that improvements would flow through to other projects undertaken by the companies involved.

The report recognized the principal contractor would have primary control of the construction site. However, an argument was made that Government agencies should attempt to influence WHS by designing a tendering process to ensure contractors are allocated Government work only if they have good safety records and the capacity to complete a project as safely as can be reasonably expected.

At the time *Getting Home Safely* was written, the ACT Government used third party certification as part of the pre-tender process. Under this approach, eligibility to tender for Government construction work required contractors to be prequalified. Prequalification included auditing and accrediting a contractor's health and safety management system. However, the report noted that once a contract was awarded, Government agencies' role in WHS became largely passive. The report also noted that although third-party accreditors were overseen by an overarching assurance body, they were paid for their work by the construction companies being assessed. Potentially, this called into question the willingness of assessors to objectively assess their clients. A need was identified for Government agencies to proactively oversee WHS following the tender process and commencement of construction projects.

Getting Home Safely recommended implementing a new Active Certification Program for construction procurement. Under this scheme, the ACT Government would employ its own auditors to conduct regular, ad hoc audits on Government-procured construction projects. These audits would include field-based assessments to check that work practices measured up to standards of performance documented in construction companies' WHS policies and procedures. Also, deficiencies identified through audits would attract demerit points, with accumulation of 100 points resulting in immediate prequalification suspension, with a review after 3 months. Significant deficiencies could

(Continued)

Case Study 2.3 (Continued)

also be referred to WorkSafe ACT for investigation and enforcement action as appropriate, and/or to the client Government Directorate to consider whether the contractor should be served with a 'show cause' notice for possible termination of their current contract.

Getting Home Safely also recommended changes to the way the safety capacity of companies tendering for Government construction projects be assessed. The original approach was to determine whether a contractor met or did not meet WHS prequalification requirements. The report argued this approach discourages construction companies from doing more than meeting bare minimum requirements to demonstrate compliance. An alternative comparative assessment of tenderers' WHS approaches and past performance was recommended. Under such comparative assessment, safety and other factors, including price, would be weighted and comparatively assessed. It was acknowledged this would not necessarily result in the best WHS performer winning a tender, but good WHS performance would give construction companies a competitive advantage in winning work. The weighting placed on WHS selection criterion would also play an important part in determining tender outcomes.

Getting Home Safely suggested a minimum threshold may need to be established for weighting the safety criterion to ensure poor safety performers did not win tenders because their performance on other criteria was sufficiently high to outweigh any WHS deficiencies. Over time, this threshold may be raised as the construction industry's performance improves.

The report also suggested Government agencies consider withholding a percentage of the final contract price for major works, paying it on completion subject to the contractor meeting certain WHS requirements.

2.5 WHS and Price Competition

It is sometimes argued that the competitive nature of the construction industry, and the attention paid to project cost as a selection criterion, encourages contractors to reduce costs so far as possible. The pressure to reduce costs is driven further along the supply chain and can have adverse WHS impacts (Manu et al. 2013; Mayhew et al. 1997). The MCF suggests price should not be the sole criterion for selecting a service provider.

In Australia, the WHS manager of a major construction contractor was interviewed during a study of client WHS activity, and described how:

> … some clients will allow us to choose best value, which means that if you can demonstrate that [a subcontractor] is safer, that they're more environmentally friendly and they're community conscious, that they add value to the end product through quality, then our clients are prepared to pay more. But some clients make the decision when it's under a certain procurement route they will choose the subcontractor that is the cheapest and expect us just to manage that. Now that's not promoting the safest option. That's promoting the cheapest option.

Cost overruns often occur in projects that are significantly behind schedule, with the Perth Arena in Western Australia and Edinburgh Trams (Scotland, UK) projects being high profile examples (see: Murphy 2010; Railnews 2012). Ball (2014) claims clients are often dissatisfied because construction projects frequently take too long, cost too much, do not meet the user requirements, fail to last the design life, or require extensive remedial work. According to Atkinson (1999), time and costs calculated in a project's early stages are guesses at best, as insufficient information is known about a project when it is costed (Ahiaga-Dagbui and Smith 2014). In Australia, the Australian Capital Territory Government (ACT 2012) observed that some public sector clients make unreasonable demands relating to project completion times which, combined with the threat of financial penalties, can have detrimental impacts on WHS performance (see, for example: Hinze 1997; Goldenhar et al. 2003; Lingard and Rowlinson 2005; Seo 2005; Mitropoulos and Cupido 2009; Oswald et al. 2013; Han et al. 2014). Reason (2008) explains that companies must obey both the ALARP principle (keep risks 'As Low As Reasonably Practicable') and the ASSIB principle (and still stay in business). Pressures to minimize costs and yet still deliver projects to tight schedules can lead to an escalation of WHS risk which can sometimes be traced back to price competition at the tendering stage. This was the case on a large construction project in the UK 2016 at which significant production pressure and cost-saving strategies that increased WHS risk could be traced back to awarding the project to an organization submitting an initial low bid (Oswald, 2016). The cost-saving strategies included delivering the project with insufficient labour and resourcing, employing migrant workers with no formal translators, and providing poor quality temporary structures, machinery, and equipment. The potential reputational damage that poor WHS performance can have for client organizations, as well as contractors, should encourage them to regard WHS as an integral component of delivering project value.

In some instances, attempts have been made to remove WHS from price competition, of which the Pay for Safety Scheme (PFSS) in Hong Kong is perhaps the most notable example. The Hong Kong SAR Government launched the PFSS in 1996 (Chan et al. 2010; Choi et al. 2012). Under the PFSS, pricing for safety-related items is removed from the competitive bidding process. Approximately 2% of the total contract sum is reserved for safety-related items. However, the Hong Kong Government Environment, Transport and Works Bureau (ETWB) explains that:

> Notwithstanding the general rule that the total value of safety items is set at about 2% of the estimated contract sum/total estimated expenditure, the price for each item should be realistic even if this means exceeding the 2% guidance.
>
> *(Environment Transport and Works Bureau 2000)*

In conjunction with the PFSS, an Independent Safety Auditing Scheme (ISAS) audits and certifies contractors' safety performance. Payment is made only if the contractors comply with a list of site safety items and receive certification for payment. Typical site safety items specified under PFSS include (Chan et al. 2010):

- developing a project safety plan;
- providing a project safety officer;

- attendance by managers at site safety committee meetings;
- occurrence of weekly safety walks;
- providing trade-specific advanced safety training;
- providing induction and toolbox training; and
- participation in safety promotional campaigns as instructed by the client's representative.

Choi et al. (2011) identified a number of benefits flowing from the PFSS. Most notably it is reported to enhance the WHS climate and attitude, promote effective WHS-related communication, streamline WHS procedures, and ensure adequate WHS training. However, the scheme has also been criticized as overly bureaucratic and costly to implement (Choi et al. 2012).

Although the PFSS seems to have produced improvements in safety performance in the Hong Kong SAR's construction industry, discussion with senior industry representatives about the potential to pay contractors for WHS performance in other jurisdictions suggests this approach is not favoured. A UK-based WHS director explained that, in his opinion, paying for WHS suggests it is an optional extra:

> So we used the argument that you didn't need to pay extra. I mean in a way it would be a bit like saying, "Can I have two prices?" I'd like a price for that scaffolding, and I'd like to know how much extra you'd like to be paid so it doesn't collapse while we're using it.

A commercial director, also from the UK, explained he was opposed to linking contractor payments to satisfying WHS requirements 'because you can't drive health and safety by money'.

Another (Australian) commercial manager questioned the need for financial drivers of WHS:

> I don't need a commercial incentive for that. Is that not enough incentive for me? I don't need any more incentive. I don't need… extra dollars or the potential of saving dollars. Yes, everyone knows that a fatality or a severe incident costs money. But we shouldn't need a commercial incentive to make the world a safer place to live.

2.6 Project Commercial Frameworks and WHS

Other recommended client actions for driving WHS performance are including WHS in project contract documents and specifying performance targets and KPIs.

The remaining sections of this chapter describe a research project that explored the impact of the project commercial framework on WHS in construction projects. The commercial framework used to deliver a construction project is designed by the client. It incorporates the set of commercial strategies and practices the client uses to establish commercial relationships to achieve their ultimate goals and objectives in a construction project. This definition is deliberately broad and encompasses:

- the contracting strategy;
- establishing project objectives and metrics to evaluate performance; and
- financial incentive mechanisms applied to the project.

Interviews were conducted with 32 participants who were either client or contractor representatives engaged in delivering large infrastructure projects. Participants were WHS managers (34%), commercial/financial managers (25.0%), project managers (31.3%), and other roles (9.4%).

The research revealed that large government clients actively design project commercial frameworks in an attempt to drive exceptional WHS performance. However, several important points of difference between clients' intentions and contractors' experiences were observed.

2.6.1 Choice of Contracting Strategy

Interview participants had direct experience of working in projects procured using design and construct (D&C), as well as more collaborative contracting strategies (alliances and delivery partnership arrangements). Generally speaking, client representatives were of the opinion that WHS performance is more readily achieved under collaborative contracting arrangements. One client WHS director commented: 'If you sum the performance of the alliances, the alliances performed basically better… more than 100 per cent better than the D&Cs in the safety space.' Another client project director reflected: 'I don't think I've been involved in one alliance where there hasn't been a positive [WHS] outcome.'

In contrast, contractor representatives were less likely to attribute high levels of WHS performance to the particular contracting strategy selected for a project. A contractor project manager commented:

> Ninety-five to 98 per cent of people on the project really perform their day to day work the same as whether it's an alliance or a D&C. They're just out there to do the best they can, and build things as efficiently and quickly and safely as they can, so… I didn't see any performance difference in the behaviours of people associated with health and safety or delivery.

A recurring theme in contractors' discussion of factors impacting WHS performance was the role of project culture. An alliance general manager explained:

> I think we had an exemplary safety performance… I'll be honest, I think it could have been any form of [contracting strategy], we could have got the same results. I'm not sure the fact that it was an alliance did anything to it… I think the things we did to turn safety performance around, you could have done just as easily on a D&C job. It was more about hearts and minds and basics than using any

collaborative mechanisms… It's more of a cultural regime and safety is just one of the outcomes.

A contractor commercial manager similarly commented: '… no matter how you procure something, this is culture… that you can drive, and as a contractor we drive that on every single project, no matter what it is. No matter what sector it is, no matter what procurement route it is.'

The reference to culture was echoed by client representatives who described how aspects of the project culture, particularly relating to collaboration, communication, and developing a shared commitment to WHS, were important drivers of WHS performance. Both client and contractor representatives suggested contractual conditions and relationships play a role in defining the project culture. A client WHS director observed:

> If you have a commercial framework that is old fashioned, based upon saving up those variation claims for a kind of a legal bun fight at the end of the project… if you have people being bullied every time there is a project performance meeting… if you're using the contract to wag the finger at the contractor… if you're wielding penalties rather than incentives… if the whole approach [to] the contract is once it is set in stone [it] is designed to drive contractors into a corner and squeeze performance out of them… if that is the way that you operate your contractual arrangements, health and safety will suffer and you won't get the best out of the contractor.

Another client project director similarly explained:

> Generally, it's not really the commercial framework that makes that big a difference… [but] because we all share the risk under that model [alliancing], and the way it's structured with the client owner in the organisation structure, it creates a much greater collaboration between the parties, and a greater culture… So it's more about, "Look we're all working together to make it work," which then drives a really strong team culture.

Relationships and team chemistry were seen as critical to the effectiveness of project performance and WHS. A contractor commercial manager explained: 'You can make a huge impact on safety no matter what the commercial framework is. Because it's people generally who are the solution to how we get better at things.'

An alliance general manager expressed a similar view:

> I've worked across all the different procurement types with the same client and it will come back to the relationship of the team in the contractor side and the team in the client side. You can have a difficult D&C type job that two very positive and forward looking proactive teams who understand each other's goals will make a success of it. Whereas, if you replace one of those parties… it'll become a big disaster, the same job. I'm firmly of that belief, if [the] client understands what [the] contractor needs to get out of it and [the] contractor understands what [the] client needs to get out of it, then you're halfway to having success.

Although contractors were less likely to attribute WHS performance to the client's choice of contracting strategy, clients perceived that a collaborative project culture was more likely to develop when they are more actively involved. A client WHS director explained: 'The commercial framework wasn't the visible driver of health and safety, it created the context. It gave us the room within which we could forge those partnerships and maintain that conversation through the works.'

Clients also recognized the opportunity, in a collaborative contracting strategy, for clients to be involved early in discussions about how WHS is to be managed and paid for in a project. A client project director explained:

> I think as soon as you put some money on the table, the alliances will put some resources on the table and focus. You take the money off the table... the response is simply, "that's all the money we've got in the budget. That's all we can afford to put in to win the job..." That's what you've got to work with... D&C is the opposite. The observations we had of the D&Cs were because they're hard dollar fixed price, and the contractor is trying to make as much money as possible, we had the minimum sized safety teams they could get away with.

Similarly, a contractor project director explained how:

> The one thing I would say about an alliance is that, certainly during the negotiation period with clients, that when you're developing a TOC [Target Outturn Cost] or a budget, there is a little bit more... respect from a client that you actually do allocate certain money for training and safety culture programs, apart from just the normal day-to-day safety stuff... And I think clients, especially government departments, when they have a bit of skin in the game on the actual overall TOC, are prepared to acknowledge that some money is needed to be spent on that.

Participants described changes to the operation of collaborative contracting approaches (in particular alliancing) in recent years. In early forms of alliancing, price competition was not a major factor in selecting a consortium to deliver infrastructure projects. Selection decisions were based on a delivery team's cultural alignment and capability to deliver a project. However, in the state of Victoria, the Department of Treasury and Finance (DTF) completed a review of collaboratively delivered construction projects. It found that, although owner representatives rated project performance in areas of non-price objectives (including WHS) as being above expectations, there was little indication that outstanding outcomes were actually achieved in collaboratively procured projects (Department of Treasury and Finance 2009). The DTF review recommended using competitive processes as a default position in the procurement of public construction projects, with one of the key selection criteria being price, unless compelling reasons for non-price competition are identified and approved. Subsequent to this review, a new form of alliancing has emerged in which two consortia are selected based on a suite of non-price criteria. These consortia are then invited to submit prices for the project and a decision is made based on cost. Tamburro and Wood (2014) argue this approach permits a highly collaborative form of delivery for projects with undimensionable risk, but also helps ensure value for money in a way earlier forms of alliancing did not permit.

Contractor representatives perceived that this approach undermines the opportunities to achieve high levels of WHS. An alliance general manager explained:

> … now is the opportunity for that commercial framework to be finalised. They [the client] haven't signed their final agreement, they've got a development agreement and they have every opportunity to set the framework to drive performance… You'd set the framework out in the first place and two companies would bid against it, I would say you would lose your opportunity to set that exceptional performance… Don't get me wrong. I'm not trying to say that organisations will only go for good [performance] if there's something in it… but I think if you wanted to go above and beyond and go really into a new space, you're down to one organisation working with one alliance team, or one consortium working with the client to get to that point… By doing competitive TOCs you essentially take away that whole principle of having an alliance in the first place. You're essentially just having a fluffy D&C.

2.6.2   Financial Incentive Mechanisms

The interviews revealed that the way financial incentives have been linked to WHS performance and applied in delivering large publicly funded projects has changed over time. Previously, positive financial incentives were provided if levels of performance above minimum conditions of satisfaction (MCOS) were achieved. These financial incentives were based on performance measured using an index of leading and lagging performance indicators (see also Chapter 6).

However, the use of positive incentives has ceased, as explained by a client WHS manager:

> Safety is simply something that they are legally obliged to achieve anyway, and the law's very clear about so far as reasonably practical, we don't actually pay positive money for safety, we only take money away.

Thus, currently used commercial frameworks incorporate a loss of potential gain-share in the event of poor WHS performance, but do not provide for positive payments if high levels of WHS performance are attained.

The financial consequences of these provisions can be severe. A contractor project manager described how, '… if we have two major safety issues, then we lose all our profit margin on this job'.

A client project director explained:

> In the more recent commercial frameworks, where we've got the trapdoor, you drop through if there's a poor safety outcome. So part of the rationale there is, if the project's really successful you should earn better than your normal margin because it's been successful. So you've delivered ahead of time, you've delivered better than the expected quality, you've delivered less disruption… you should get a better return. But there's an incongruence that if we do all those things and someone gets hurt or there's a serious safety incident, that you should

be getting a bonus when someone's hurt. So that's why the safety part of the commercial framework's like a trapdoor, so it does actually penalise significantly the return that the organisation can make. So 50 per cent for instance, or 25 per cent, is still a significant fee reduction.

However, applying negative incentives was also recognized as creating a challenge in the event that a serious WHS incident should occur early in the life of a project. A client project director explained how in recent projects:

> There was no upside, or no gain I suppose, or commercial gain, by achieving the safety objectives. But there was pain if you dropped below certain objectives… which then led to conversations around what do we do to ensure we're above minimum conditions of satisfaction for safety performance, and also, how do we make sure the other related key result areas ultimately can help us if something actually goes wrong, early days in the commercial framework?… We set up the commercial framework so that if, for some reason, there was a really poor safety outcome very early, that it just didn't kill the whole job culturally… and the commercial framework provided an opportunity for people to recover, and still focus on the things that are important… Because if you have one failure in safety and there's no other benefit to continue to refocus on safety, then will you?

Thus, a so-called 'claw back' provision is incorporated which provides contractors with

> … an ability to claw back. And the rationale behind that is things happen, not that we should be that fatalistic, but if there was an incident that happened right at the beginning of the project, you don't want to doom the project to never be able to get out of the doldrums. So the idea is to still provide some incentive, and the claw back is about really the team having some sort of proactive safety program that actually drives better safety outcomes. So if the incident happened at day one, for instance, you can still get back to some level of performance. You'll never get back to 100 per cent, but there's still an incentive to drive better than normal safety performance because, you know, there's an ability to make a return; it's not dropped away forever.
>
> *(client project director)*

Contractor representatives perceived the use of negative financial incentives as damaging to collaborative working relationships between clients and contractors in projects. One contractor WHS manager explained:

> Negative incentives don't create a collaborative environment to resolve key issues. So it may push through [and] allow the approach to get over the line. But… the negative incentives don't allow for collaboration and promoting innovative approaches, whereas positive incentives work towards getting clients and contractors working together to demonstrate value for money, and that adds value to the client for future jobs.

Similarly, the general manager of a collaborative alliance project believed penalties do not effectively motivate people to strive for exceptional WHS performance:

> I think all the arrangements are seen to have been quite punitive. I have seen a number of occasions where that punitive side has come out and there's little incentive to do anything different other than your moral obligation. Afterwards, you kind of cut back and you've earned all your dollars... So if you were driven commercially alone, you just wouldn't put the effort in you were before, but your moral obligation... I'm not sure. I think it's a lot more stick than carrots.

The negative implications associated with penalizing poor WHS performance, without recognizing or rewarding positive performance, were further explained by another contractor WHS manager:

> We've had conversations, commercial conversations, and although clients have been reluctant to remove the old pain injury rate measurements, we've added [positive performance] on the indicators. Where they keep them as pain, they don't introduce them as a gain. So it's all if you don't perform, if you don't meet this target, then... you're penalised this money. So it's a negative conversation.

Gudiene et al. (2014) found client establishment of clear and precise objectives is a critical factor for project success. However, the research suggested negative financial incentives potentially reduce collaboration and can also produce unforeseen and undesirable consequences (Kadefors 2004). Clients recognized that, in some cases, contractors simply price the risk of a financial penalty into their tender submissions. As a client project manager explained:

> We saw in the submissions that by leaving the penalty at the larger level, we actually paid for it. Because [the contractor] said, you know, 'If you leave it at that, we'll cost the risk of something happening', because they needed to go through their risk analysis. 'If you bring it back to what you had before, so essentially halve it for this contract, we can offer you a saving of about a million dollars.' So the client's always paying for that risk.

Both clients and contractors also acknowledged that linking negative incentives to lagging WHS indicators could encourage underreporting and reduce the reliability of WHS performance data. This problem is also discussed in Chapter 6.

Based upon the interview data collected in this research project, Figure 2.2 identifies some challenges and considerations for client organizations seeking to use commercial frameworks to influence WHS performance.

2.7 The Potential for Unintended Consequences of Client WHS Activity

In making significant policy changes, governments can have a substantial influence on the way private sector organizations practise WHS. But there is a risk that clients'

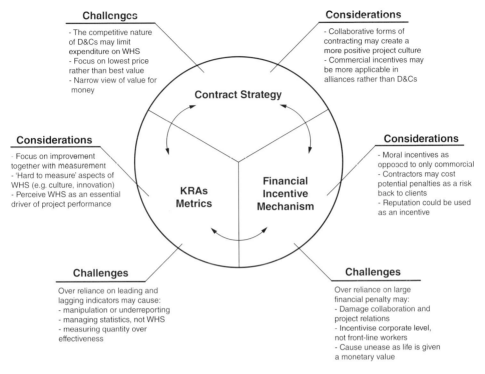

Figure 2.2 Considerations and challenges inherent in designing a commercial framework.

expectations regarding WHS may, in some instances, produce unintended consequences. For example, interview data collected from construction contracting organizations in Australia reveals that increasing proceduralization of WHS is sometimes unhelpful because it creates onerous paperwork without necessarily improving the state of WHS. As the project manager at one large Australian infrastructure project explained:

> Clients predominately in the government sector are perhaps still a few years behind... Why are they behind? Very procedural driven... If I have an incident, then there's a 45 day period to return our report that's got to be 25 pages thick, it's got to have six appendices... But in reality, is that the right approach to looking at an incident? Nine times out of ten with incidents, it doesn't take very long to get to the root cause, and then we spend a lot of time fussing about trying to put that into some sort of document that fits a template that fits an expectation... Government clients are very procedurally driven... everyone has a policy or a procedure for something.

He went on to describe how private sector contracting organizations have become less focused on bureaucratic aspects of documented WHS systems, preferring to emphasize flexibility and workforce engagement:

> I think that the construction world has evolved from [a procedure-driven approach]. So rather than having 400 procedures to execute their business and

they audit themselves to death over whether they're following every procedure, it's become a more dynamic process. So it's more headline: these are the important things. It's taking the procedural side out and putting it back to policy, if you like, and just saying: 'look these are the givens, these are the things that we must strive to achieve or mustn't do.' But we let the procedural side of it sort of evolve into a more positive engagement with the way that we conduct our business.

Rules and procedures specify how a work process, task, or activity should be undertaken. They are seen as essential in directing, standardizing, and monitoring work (Hale and Borys 2013a). However, it is also recognized that documented procedures cannot cover all eventualities, and are often developed at a general level of abstraction by people who may not understand the applicability of rules or procedures to local conditions (Iszatt-White 2007). Hollnagel (2015) observes inevitable gaps between work as imagined (by managers and technical WHS specialists) and work as done (by workers responding to localized conditions and circumstances). Fucks and Dien (2013, p. 32) similarly warn that growing bureaucratization of WHS can reduce people to the status of 'robots' whose unthinking compliance with rules takes precedence over situational awareness and responsiveness to emergent danger in the work environment.

A growing culture of resistance to interpreting and applying modern WHS legislation has been observed in the construction industry (Waddick 2010). In an ethnographic study of workers in the Scottish construction industry, Oswald (2016) notes that WHS-related rules are often seen as being inflexible, inappropriate, and sometimes unrealistic in the context of environmental constraints and work schedules. Löfstedt (2011) suggests the problem lies less with the regulations themselves and more with the way they are applied, as many WHS regulations have been misinterpreted or misapplied. The ACT Government's *Getting Home Safely* report (2012) noted that the wording of WHS legislation is sometimes difficult to comprehend, and that related codes of practice would be more useful if they were shorter and more practically based. Clients, alongside contractors, play an important role in interpreting government policy to make their WHS policies and procedures comprehensible and appropriate. The role of formal rules and procedures in the management of WHS is further explored in Chapter 7.

Bieder and Bourrier (2013) also warn that critical information required by workers is buried inside long, overly complicated documents. The ACT Government (2012) recognized a need to minimize unnecessary paperwork and clarify safe work method requirements; to shift the emphasis from paperwork to safe work practices. This is particularly important considering 'safety isn't an entity – something separated from work and practice and sat on its own in a folder on the shelf of a site cabin – but something we can create between ourselves on sites on a daily basis' (Sherratt 2016, p. 181). A contractor commercial director in the Australian construction industry described how long and complicated WHS procedures can make important WHS information inaccessible to workers:

Each risk assessment and method statement are developing into 100 pages long... and the problem is that the guy on the drill, who's actually building the job, is never going to read 100 pages. The only two bits of paper you need, the guy wants the two bits just to tell him how to do the job safely, and invariably that's two pages out of that 100 pages. So what we've tried to do here is to streamline method statements.

2.8   The Overriding Importance of Relationships

Trust can be eroded when clients are perceived to place too much emphasis on 'policing' contractors' compliance with WHS rules. A client WHS manager described how '... you've got to back the safety police off and empower the blokes [workers]'.

The above comment suggests that clients tread a fine line. At one case study construction project, the importance of maintaining a good client-contractor relationship was observed by the WHS manager. Contractors at this project were required to document how their work processes complied with a framework document developed by the client. Part way through the project the WHS manager observed contractors were becoming disengaged. He explained:

> The morale's low. It's showing in their work. They're having more hazards and risks than ever, and if we keep on applying this policeman approach, it's going to really fester. So we started to try and figure out what we could do. And from a safety space we started working a lot closer with them – their safety manager, and their safety team.

This altered approach led to improvement, as the WHS manager explained: 'The relationship between our safety team and their safety team has been better than it ever has. There is a lot of trust there now.'

Collaboration, trust, engagement, and fairness were all identified as preconditions for WHS success in the London 2012 Olympics construction (Bolt et al. 2012). Strong trust between clients and contractors provides a strong platform for effective collaboration through shared understanding, agreement, and commitment to project WHS objectives. Winkler (2006) similarly reported that collaborative partnering relationships between clients and contractors produced a joint approach to WHS that led to an increased focus on WHS and improved outcomes.

2.9   Conclusions

As the initiators of projects and purchasers of services, clients can play an important role in promoting and enabling WHS performance on construction projects. Clients make key decisions about how projects are procured, as well as determining budgets and timelines. Research demonstrates that the best outcomes are achieved when WHS is considered at the early project stages. Clients are uniquely positioned to make sure WHS is integrated into all project decision making, and that WHS risks are systematically identified, managed, and communicated to participants as projects progress through the stages of planning, design, construction, and completion.

Clients are increasingly interested in WHS, recognizing that poor WHS performance reflects badly on all parties involved in a construction project. Guidance is available to help clients to embed WHS into their procurement and project management processes. Some clients have even incorporated WHS into the commercial frameworks used to deliver projects. Various models have been used for specifying WHS requirements or target performance levels in contracts, and linking these to performance measurement and payment. Most importantly, research suggests strong, positive, and collaborative relationships between clients and contractors produce good WHS outcomes. Integrated forms of contracting, such as alliancing, are more conducive to achieving a unified vision and commitment to WHS between clients and other project contributors. However, good WHS outcomes can also be realized under more traditional project procurement arrangements if clients are actively engaged and collaborative.

Discussion and Review Questions

1   How much should clients be engaged and involved in WHS in delivering projects they procure? What are the constraints and dangers inherent in becoming an active client in relation to WHS and how might these be overcome?

2   How important are collaborative and good relationships between clients and contractors in delivering high levels of WHS performance? What client or contractor behaviours can facilitate or impede the development and maintenance of trust in contractual relationships?

3   Can WHS be improved by including it in the commercial framework used to deliver a construction project? What are the challenges or potential benefits associated with clients driving WHS through commercial processes?

Acknowledgements

A large component of this chapter is drawn from a research report funded by the Major Transport Infrastructure Program, Department of Economic Development, Jobs, Transport and Resources, Victorian Government. The authors also gratefully acknowledge the support of the Office of the Federal Safety Commissioner in funding the development of the MCF. Early work on client WHS initiatives was supported by Australian Linkage Project Grant (LP120200440).

3

Designing Safe and Healthy Products and Processes

Helen Lingard, Nick Blismas, and Payam Pirzadeh

School of Property, Construction and Project Management, RMIT University, Melbourne, Victoria, Australia

3.1 Safety in Design

Professional designers have had a longstanding focus on, and responsibility for, the structural safety of the buildings and facilities they design. Yet the notion of designing for workers' health and safety is a relatively recent phenomenon which emerged from a growing belief that safety incidents in construction operations can be traced back – at least in some measure – to design decisions. The case for safety in design was evidenced by numerous analyses of historical incident data and supported by theoretical models of incident causality. As a result, considering construction workers' safety and health in the design stage of construction projects has become a key feature of construction health and safety policy. In a number of industrialized countries it is legally mandated. However, there remain significant challenges in the practical implementation of safety in design in the construction industry. This chapter:

- considers the case for considering safety in design and the resulting policy responses;
- explores the challenges and dilemmas experienced in the practical implementation of safety in design in the construction industry;
- identifies key principles and practices that may help to address these challenges; and
- explores the scope for improving safety in design practice in the future.

3.2 The Case for Safety in Design

The widespread acceptance of safety in design has grown partly from a theoretical understanding that design decisions can be a causal (or at least contributing) factor in workplace safety incidents. Probably the best-known model of incident causation that traces incidents back to a variety of organizational failures (including system design) is the 'Swiss cheese' model (Reason 1997). The model explains human error as being caused by various 'upstream' systems failures rather than by the idiosyncratic nature of people's behaviours and beliefs.

Integrating Work Health and Safety into Construction Project Management, First Edition.
Helen Lingard and Ron Wakefield.
© 2019 John Wiley & Sons Ltd. Published 2019 by John Wiley & Sons Ltd.

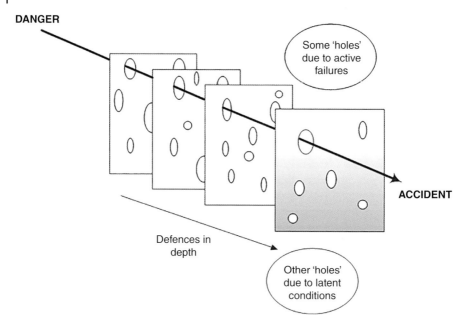

DANGER

Some 'holes' due to active failures

ACCIDENT

Defences in depth

Other 'holes' due to latent conditions

Figure 3.1 'Swiss cheese' model. *Source:* Reason (1997).

The 'Swiss cheese' model is depicted in Figure 3.1. It suggests a system has a number of defensive layers that prevent incidents. These defensive layers are not perfect, each having its own gaps or holes (hence the Swiss cheese analogy). These holes open and change over time. The presence of holes in any single layer does not usually result in an incident because other protective layers are intact and serve to prevent an incident. However, sometimes, the holes line up, providing the trajectory or pathway through which an incident can occur. The model differentiates between active failures and latent failures.

Active failures are unsafe acts, such as mistakes, slips, lapses, or even deliberate rule violations. Active failures are present in many incidents and are often focused upon as the cause of an incident. However, usually they can be traced back to more fundamental failures in a system.

In contrast, latent failures arise at a managerial or organizational level. They include erroneous decisions made by designers of workplaces and processes and those who establish work procedures and rules. According to Reason, latent failures can create conditions that produce errors or encourage rule violations within a workplace (for example, in creating time pressure, deploying inadequate equipment, or providing insufficient instruction and training). But latent failures can also create enduring weaknesses in a system's defences against incidents. These weaknesses can lie dormant for long periods and may not be recognized until an incident occurs.

The 'Swiss cheese' model has been used to understand safety incidents occurring in the construction industry. For example, Priemus and Ale (2010) investigated the Bos and Lommerplein estate project in Amsterdam. They adopted the latent failure concept to explain how systemic barriers in design, construction, permitting, inspection, and use stages resulted in serious structural safety problems. Reason's ideas also inspired a number of construction-specific incident causation models. Manu et al. (2010) identify various proximal and distal factors contributing to incidents in construction, suggesting that many of these characteristics can be traced back to the clients' brief, design decisions, and project management decisions.

A research team at Loughborough University (UK) developed the Construction Accident Causality (ConAC) model (Haslam et al. 2003). ConAC was based on analysis of 100 safety incidents in the UK construction industry. The research team used the information obtained from people involved in selected incidents, including the injured workers and their supervisors, to describe the processes of accident causation in construction. The resulting model, depicted in Figure 3.2, identified originating influences (akin to latent failures) as:

- client requirements;
- features of the economic climate;
- prevailing level of construction education;
- design of the permanent works;
- project management issues;
- construction processes;
- the prevailing safety culture; and
- the risk management approach.

Researchers in the USA (Behm and Schneller 2013) and Australia (Cooke and Lingard 2011) have used the ConAC model to analyse the causal factors in construction incidents. Cooke and Lingard drew on coronial findings to analyse 258 work-related deaths arising from injury in the construction industry and occurring between 2000 and 2010. In the USA, Behm and Schneller (2013) investigated 27 construction accidents using the ConAC framework as a guide during the investigation process, which included interviews with various employees, supervisors, managers, and safety representatives. In these analyses, the originating influences, shaping factors, and immediate circumstances encompassed in the ConAC model were all identified, although their relative importance varied (Gibb et al. 2014).

3.3 How Important Is Design as a Cause of Construction Incidents?

A number of studies have analysed previous incidents in an attempt to quantify the link between design and construction workers' safety. Behm (2005) reviewed 224 fatality reports from the Fatality Assessment Control and Evaluation (FACE) database held by the USA's National Institute for Occupational Safety and Health (NIOSH). He considered an incident to be design related if one or more of the following criteria were met:

- the permanent features of the construction project were a causal factor in the incident; and/or
- any of the design suggestions identified in previous studies could have been implemented to prevent the incident; and/or
- modification of the design or the design process could have prevented the incident.

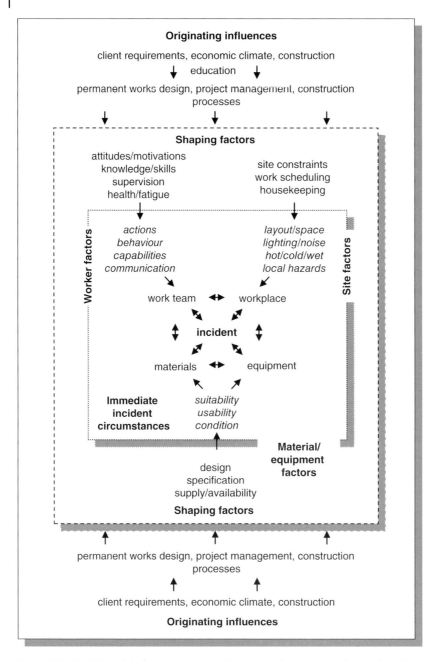

Figure 3.2 ConAC model of construction accident causation. *Source:* adapted from Haslam et al. (2003).

Behm reported design to be a causal factor in 42% of fatal accidents reviewed. He concluded that safety in design can:

(i) positively affect the safety of construction workers during construction work, as well as in a facility's subsequent maintenance, renovation, and repair work; and

(ii) reduce safety risk across all types of construction projects.

The results of Behm's (2005) study were validated by Gambatese et al. (2008). They asked an expert panel, composed of construction industry professionals, to review a subset of the 224 fatality cases. The panel judged whether the design was a contributing factor to the incident. The panel considered there was a link between the incident and the design if:

- the permanent features of the project could have been modified to prevent or reduce the risk; and/or
- the construction plans and specifications could have been prepared in a different way to avoid the incident; and/or
- the construction safety risks related to the design could have been communicated to the constructor to avoid the accident.

In 71% of the fatalities investigated, the panel's responses confirmed Behm's findings of a significant link between the design and the incident (Gambatese et al. 2008). In Australia, Driscoll et al. (2008) reported that 44% of construction fatalities were 'design related'.

Thus, taking the international evidence into consideration, design seems to be a factor in somewhere between 40% and 50% of incidents. However, is also likely that researchers who go looking for design as a contributing cause will overestimate the extent to which it is found. Lundberg et al. (2009) referred to this tendency as 'What You Look For is What You Find'.

Hale et al. (2007) also point out that research into safety in design that is based on the analysis of past incidents is subject to large differences of interpretation. This was alluded to by Driscoll et al. (2008) who qualified their research, stating that 'informational difficulties' made it difficult to ascertain whether the construction fatalities they analysed could be attributed to:

- the permanent design of the building/structure;
- the design of plant/equipment; or
- the design of the process of construction, including temporary works.

Driscoll et al. referred to limitations in the quality of data available to them. However, 'what is being designed?' is a question which creates boundary problems that have a real impact on how safety in design should be understood and operationalized in the construction industry context.

At present, it is entirely unclear when design work commences and where it finishes in the life of a construction project. Does a client's early definition of the need for a project, and establishment of specifications for a facility, constitute design work? Or does design work commence when a design consultant is engaged to develop a facility's conceptual or detailed technical design?

A particularly problematic question is, 'at what point does design end?' It is problematic because construction contractors make multiple changes to design as work progresses. Should designers consider the processes of construction, maintenance, and operation of a facility? As Kinnersley and Roelen (2007) point out, in aviation the operating procedures for a plane are considered part of the design. If designers are responsible for designing for safety in construction operation and end use, how can they control for situations in which a design is changed, and/or people construct or use the facility in ways that they did not foresee?

We will return to these boundary problems later in the chapter when we discuss the challenges inherent in the practical implementation of safety in design in construction.

3.4 The Policy Response

In response to emerging evidence of a link between design and workers' safety, policy makers and legislators have formalized the expectation that organizations address safety in design in all industries – including construction.

The first iteration of the Construction (Design and Management) Regulations was introduced in the UK in 1994. These CDM Regulations established specific statutory work health and safety (WHS) duties for clients and designers. They also required the creation of a project-specific health and safety file to ensure health and safety information was documented and communicated through all stages of the project lifecycle. Since then, the Regulations have been subject to numerous reviews and changes. A new version of the Construction (Design and Management) Regulations came into force in 2015, revoking and replacing older versions (The Construction (Design and Management) Regulations 2015 (UK) 2015). Changes to pre-construction phase activities, required under older versions of the Regulations, were made in response to perceptions that the requirements were very bureaucratic, costly to implement, not well integrated into existing construction project management processes, and provided little benefit for improvements to workers' health and safety (Beal 2007).

The Construction (Design and Management) Regulations 2015 were simplified and restructured to better reflect construction project processes. The client was given additional responsibilities, including a number of responsibilities originally allocated to the role of a CDM Coordinator under the 2007 version of the CDM Regulations (Construction (Design and Management) Regulations 2007 (UK) 2007). Clients are now required to ensure suitable arrangements are made for effective communication and coordination between members of the project team, and to ensure the principal designer (PD) and principal contractor (PC) carry out their duties effectively. The role of CDM Coordinator no longer exists. The occupant of this artificially created role was previously responsible for coordinating the health and safety-related activities of multiple project participants. Under CDM 2015, many of the duties previously assigned to the CDM Coordinator have been transferred to the PD whose primary role is to manage the pre-construction phase of the project. This includes:

- planning, managing, and monitoring the pre-construction phase, and coordinating matters relating to health and safety, to ensure that, so far as is reasonably practicable, the project is carried out without risks to health or safety;

- identifying and eliminating or controlling, so far as is reasonably practicable, foreseeable risks to the health or safety of any person:

  (a) carrying out or liable to be affected by construction work;
  (b) maintaining or cleaning a structure; or
  (c) using a structure designed as a workplace;

- ensuring all designers engaged in the project comply with their duties;
- assisting the client in obtaining and providing appropriate pre-construction phase information, and providing this to designers and contractors appointed to the project;
- liaising with the principal contractor throughout the duration of the principal designer's appointment and sharing information relevant to the planning, management, monitoring, and coordination of health and safety matters during the construction phase; and
- preparing the project health and safety file, to be handed to the client at the end of the project (Health and Safety Executive 2015).

More than a decade after the UK first introduced legislation requiring designers to address the health and safety of construction workers, Australian legislators followed suit. The State of Victoria was an early adopter of the regulatory approach, introducing a responsibility for designers of buildings or structures in the Occupational Health and Safety Act 2004 (Vic.) (2004; s.28) that came into force in 2006. In contrast to the UK approach, the Victorian Act requires that a designer of buildings or structures

> who knows, or ought reasonably to know, that the building or structure is to be used as a workplace must ensure, so far as is reasonably practicable, that it is designed to be safe and without risks to the health of persons using it as a workplace for a purpose for which it was designed.
>
> *(Section 28(1) of the Act)*

However, the Victorian Government guidelines suggest this duty does not extend to consideration of the health and safety of construction workers, stating 'the duty does not include the design of the construction and demolition phases of a building or structure's lifecycle', and adds as a footnote that, 'During construction and demolition, the building or structure is not being used as "a workplace for a purpose for which it was designed"' (WorkSafe Victoria 2005, p. 3).

A subsequent attempt to harmonize occupational health and safety legislation across all Australian states and territories saw the introduction of a Model Work Health and Safety Act and Model Work Health and Safety Regulations. The Model Act and Model Regulations included responsibilities for designers of buildings and other structures. The Model Act requires designers of structures to be used as, or at a workplace to ensure, so far as is reasonably practicable, that the structure is designed to be without risks to the health and safety of persons who:

- use the structure for a purpose for which it was designed;
- construct the structure at a workplace;
- carry out any reasonably foreseeable activity in relation to the manufacture, assembly, or use of the structure for a purpose for which it was designed, or the proper demolition or disposal of the structure; or
- are at or in the vicinity of a workplace and who are exposed to the structure.

However, inconsistencies in the scope and interpretation of responsibilities for safety in design between jurisdictions persist. The Victorian position remains unchanged and at odds with the broader national interpretation and scope. This is unhelpful and highlights some of the boundary problems associated with the way safety in design is currently implemented through regulatory processes.

A different situation exists in the USA. Hazard Prevention through Design is identified as a strategic goal in the USA's National Construction Agenda for Occupational Safety and Health Research and Practice in the US Construction Industry (Centers for Disease Control and Prevention 2008). However, the USA has not introduced safety in design requirements in preventive occupational safety and health legislation. Thus, the motivation to implement safety in design in the USA's construction industry must come from clients or designers themselves.

Hale et al. (2007) consider the factors that shape the willingness of design organizations (and individual designers) to address safety in design. They suggest that safety in design can add to the cost and time required for design work, which can reduce profit margins and even decrease market share if clients select competitors with less rigorous safety in design processes. A great deal will rest on the conditions under which a company operates. It is possible that safety in design will become an essential part of 'doing business' as construction clients' expectations grow in relation to the proactive identification and resolution of WHS hazards before construction commences. Ethical considerations, and concern for a design organization's professional reputation, may also be motivating factors. Hale et al. (2007) suggest concerns about the liability of a company in the case of safety incidents should encourage design organizations to adopt 'state of the art' safety in design processes. Yet, in the USA's construction industry, concerns about liability have been identified as a barrier to designers addressing construction workers' health and safety. Gambatese et al. (2005) also report that of six project criteria, USA construction design professionals ranked safety as their lowest priority. Perplexingly, concerns about legal liability were one of the main impediments to US designers' willingness to address construction workers' health and safety in their professional practice.

Hale et al. (2007) argue the benefits associated with considering safety in the design stage of products are not universally accepted, and some suppliers actively try to limit their liability by pushing decisions about safety to the user of a product. This is analogous to construction, in which designers of the permanent structure are careful not to implicate themselves in the design of the processes of construction, and construction contracts reinforce a clear distinction between product and process design (see also Toole 2005, 2007). We will return to the relationship between product and process design in Section 3.5.3 of this chapter.

3.5 Challenges Inherent in Safety in Design

Several significant challenges have been identified that impede the successful implementation of safety in design in the construction industry. These are discussed below.

3.5.1 Sociotechnical Complexity

Design work in the construction industry is a complex socio-technical activity. This presents significant boundary problems, coordination issues, and difficulties in attributing responsibility for safety in design (Lingard et al. 2011b).

Table 3.1 Sources of complexity in construction projects.

Organizational complexity	Characterized by: • a significant division of tasks, • multiple organizational units and/or hierarchical levels, • multiple specializations, and • many interdependencies between organizational elements.
Technological complexity	Characterized by: • multiple diverse inputs, outputs, tasks, or specialities, and • many interdependencies between technologies, tasks, or inputs.

Baccarini (1996) defined complexity as 'consisting of many varied inter-related parts, operationalised in terms of high levels of differentiation and interdependency' (p. 202). There are two kinds of project complexity, as outlined in Table 3.1.

Construction design work exhibits high degrees of both organizational and technological complexity. This complexity is evident in the nature of structures created to deliver a project: structures of work (relationships between collaborating parties); structures of information (knowledge transactions); and structures of governance (contractual arrangements) (Lingard et al. 2007).

Construction design teams are 'temporary, multidisciplinary and network-based organizations' (den Otter and Emmitt 2008, p. 122). Design entails a network of tasks that rely on contributions from a range of specialists and the activation of a complex 'web' of interorganizational relationships.

In this environment, design decisions cannot be viewed as the sole preserve of 'the designer' – itself, an abstract and difficult to define sociotechnical role (Lingard et al. 2012a).

Research in the Australian construction industry has highlighted this complexity, revealing how design decisions that impact construction workers' health and safety can be influenced by parties external to the construction project (see Case Example 3.1).

Case Example 3.1 Client's Customers Influence Construction Workers' Health and Safety in a Rail Construction Project

The case arose during design and construction of a suburban train station.

The original concept design involved constructing a new 'island' platform, built between two existing and fully functioning rail lines. A pedestrian footbridge was to be built, spanning the full width of the tracks. Access to the platforms from the footbridge was to be provided by stairs at either end, and in the middle of the footbridge. In accordance with disability access requirements, the original concept design included an alternative means of accessing the platform by providing a lift.

However, before the contract was awarded, an incident occurred at a similar train station. This incident involved the death of a passenger who could not be removed safely from an island platform because the ambulance trolley would not fit in the platform lift. Consequently, paramedics were forced to remove the passenger by walking over 'live' rail tracks.

(Continued)

Case Example 3.1 (Continued)

In addition to concerns about access to, and egress from station platforms, there were a number of passenger complaints about station lifts breaking down. A review of design policy for train stations led to new requirements specifying that all new stations would be installed with lifts able to accommodate a standard ambulance trolley, and that an alternative means of access in the form of a ramp would also be provided. This new policy was introduced shortly after tenders closed for the railway station project and companies that had tendered for the project were given two weeks to amend their proposals.

The contract was eventually awarded to a design and construction contractor on the basis of a proposal that included a number of changes to the original concept design. One of the main changes was the addition of a ramp for disability access. The late inclusion of a ramp in the design resulted in emergent hazards during the construction stage which were not envisaged at the tendering stage. The contractor commented:

> When we priced and sketched up [the proposed design] at tender stage, no ramp was included. We were only given two weeks. We had already put our price in and it was a last minute change by the client... No-one picked up at the time about the canopies being bisected [by the ramp].

A post-award risk assessment (involving the client, the rail operator, and design and construction contractor) was conducted once the project commenced. This risk assessment focused primarily on the health and safety of end users of the station. The risk of persons jumping over the ramp balustrading onto an adjoining canopy was identified. To address this risk, 'throw screens' were designed to be fixed to the ramp balustrading to reduce the risk of people climbing, or throwing objects, over the side. The risk assessment also identified the need to provide landings at regular intervals on the ramp to provide 'rest' areas.

The addition of the ramp, the landings, and the throw screens had a significant impact on the design and construction of the station. The sizing of columns supporting the ramp had to be changed, with some columns more than doubling in size due to the inclusion of landings and throw screens. Size increases to the platform's steel structure were also required to safely support increased loads associated with the ramp and larger columns.

As a result of these changes, construction workers' exposure to hazards associated with crane lifts was significantly increased. Additional platform components needed to be lifted into place, and the larger size of structural members reduced manoeuvrability and increased risk. The rail lines had to be closed on the days of the lifting operations. Further, the reduced clearance between the underside of platform beams, which had doubled in depth, and the ground meant that services originally planned to be connected to the underside of each beam had to be relocated due to restricted access clearances. Thus, a series of holes had to be cut into every intersecting beam for the length of the platform (approx. 100 m), to allow conduit to be installed to accommodate services. The steel beams had been fabricated without any penetrations, so in situ cutting of holes presented new hazards associated with using cutting equipment in an area that was difficult to access.

This case example shows how design decisions emerge as the interests of multiple stakeholders are considered and change as new information becomes available. Some influential stakeholders are external to the project team. Thus, in the complex socio-technical environment of a construction project, the boundary of the design process may not be sharply defined (see Fadier and De la Garza 2006), and questions about how responsibility for safety in design can be appropriately allocated over the life of a project have not been satisfactorily answered.

3.5.2   Vertical Segregation

The vertical segregation inherent in construction project supply arrangements also presents a challenge for implementing safety in design. Participants responsible for initiating, designing, producing, using, and maintaining facilities are vertically segregated. Relationships are often 'arms-length' and restrictive in terms of opportunities for information exchange.

The division between design and construction functions can hinder the development of shared project goals (Baiden and Price 2011) and negatively impact project outcomes (Love et al. 1998). But, as Atkinson and Westall (2010) also point out, vertical segregation can impede the industry's capability for effectively implementing safety in design. Donaghy's (2009) recent review of health and safety in the UK construction industry identified the separation of, and poor communication between, design and construction functions as a cause of poor safety performance.

The situation is made even more complicated because product complexity means technical health and safety knowledge often resides with specialist subcontractors or suppliers who take responsibility for the detailed design, manufacture, supply, and installation of components (Haviland 1996; Slaughter 1993). Yet, these people may not be engaged at an appropriate time to seek their input into critical design decisions as they are being made. For example, Franz et al. (2013) present case study data showing how, in comparable projects, better work health and safety outcomes are achieved when specialist contractors are involved early in design decision making. Wright et al. (2003) similarly show how design solutions to identified safety problems are often driven by building systems' manufacturers rather than by principal design consultants. Yet, these people may not be engaged at an appropriate time to seek their input into safety-relevant design decisions as they are being made.

3.5.3   Confusion Between Product and Process Design

Most definitions of safety in design imply that designers should identify and address safety issues associated with facilities, structures, processes, equipment, tools, and work systems. For example, the USA's National Institute for Occupational Health and Safety (2008, p. 108) defines 'prevention through design' as:

> … addressing occupational safety and health needs in the design process to prevent or minimize the work-related hazards and risks associated with the construction, manufacture, use, maintenance, and disposal of *facilities, materials, and equipment* (italics added).

Schulte et al. (2008, p. 115) define safety in design as:

> ... the practice of anticipating and 'designing out' potential occupational safety and health hazards and risks associated with new *processes, structures, equipment, or tools, and organizing work,* such that it takes into consideration the construction, maintenance, decommissioning, and disposal/recycling of waste material, and recognizing the business and social benefits of doing so (italics added).

The Australian Strategy for Work Health and Safety 2012-22 states: Good design can eliminate or minimise the major physical, biomechanical and psychosocial hazards and risks associated with work. Effective design of the overall system of work will take into account, for example, management practices, work processes, schedules, tasks and workstation design. (Safe Work Australia, 2012b, p.7).

Notwithstanding these inclusive definitions, the interpretation of what is being designed is often unclear. Driscoll et al. (2008) reviewed the findings of coronial investigations in Australia to determine the extent to which design was a causal factor in construction industry deaths. They found that 44% of the deaths examined were design related. However, a close assessment of the accident circumstances described by Driscoll et al. reveals that the majority of the deaths were related to the design of work processes (including temporary works and equipment being used). The design of the permanent structure was clearly implicated in only one of the deaths examined and involved a maintenance worker, working on the roof of a building, falling through a fragile skylight.

It is also apparent that many commonly cited design solutions to safety problems identified in the construction industry actually involve a redesign of the construction process, rather than altering the original design of the permanent building or structure to be constructed (see, for example, Wright et al. 2003). Design of healthy and safe work processes is a neglected area in the research on construction safety in design.

This lack of clarity is unhelpful in the construction industry because it creates confusion about who should be responsible for safety in design. Different project contributors will be involved in design decisions relating to buildings (or their component parts), equipment, work processes, and so on. When implementing safety in design it is essential to have a clear understanding about what is being designed, and who the relevant contributors to safety in design are. A principal architect may not, for example, be significantly involved in designing the construction process.

> Good design can eliminate or minimize the major physical, biomechanical and psychosocial hazards and risks associated with work. Effective design of the overall system of work will take into account, for example, management practices, work processes, schedules, tasks and workstation design.
>
> Safe Work Australia (2012, p. 7)

3.5.4 Knowledge Issues

Research suggests design professionals' knowledge about safety in design is limited. This may result from a lack of formal education about construction health and safety, or from designers' limited work experience on construction sites (Gambatese et al. 2005). Brace et al. (2009), who reviewed the causes of fatalities in the UK construction industry, wrote that:

> ... many designers still think that safety is 'nothing to do with me', although there are a small cohort who want to engage and are having difficulty doing this because they do not fully understand what good practice looks like.
>
> *(p. 12)*

Also in the UK, Donaghy (2009) proposed accrediting bodies impose a requirement that work health and safety is integrated into the education programmes of designers and others engaged in delivering construction projects. Similar suggestions have been made in the USA following a study that found almost 90% of contractors believed including work health and safety as a requirement in the education of architects and design engineers would improve the industry's health and safety performance (Gambatese et al. 2008).

Gambatese et al. (2005) also report that design professionals who have limited knowledge and experience in implementing safety in design are much more likely to perceive that safety in design will increase project costs, create schedule problems, and reduce design quality.

3.5.5 Oversimplified Assumptions

Hale et al. (2007) argue that trade-offs between safety and other design criteria (such as cost, quality, production) are an inevitable part of design decision making. However, they suggest these trade-offs are not made explicit. A similar observation was made by Lingard et al. (2013b). Guidance materials on safety in design often implicitly assume design measures that reduce health and safety risk in one stage of a product's lifecycle are beneficial (or at least have no negative impacts) for health and safety risk in other lifecycle stages. However, there is evidence to suggest trade-offs are made.

When discussing implications of using built up, compared to composite panel, roofing systems, Wright et al. (2003) foreshadow the possibility of conflict between designing for occupational health and safety in the construction and operation stages of a facility. Although composite roofing systems reduce the need for work at height during installation, they present an increased risk of falling during roof maintenance (Wright et al. 2003).

As Case Example 3.2 shows, it is possible that actions taken to reduce safety risks in the end use of a facility can increase risks experienced by construction workers.

Case Example 3.2 Fire Rating a Food Processing Facility

This case arose during design and reconstruction of a food processing facility. The plant had been partially destroyed by a fire, resulting in temporary closure. To prevent loss of employment in the area, assistance was offered to the client to support reconstruction and the planning process was fast-tracked to facilitate this. As a consequence of this support, the client decided to rebuild the plant and appointed a contractor under a design and build contract.

(Continued)

Case Example 3.2 (Continued)

The client originally requested that a sprinkler system not be installed in the food processing building. After construction work had commenced, however, a registered building surveyor advised that, if a sprinkler system was not installed, a fire-rated wall would have to be incorporated into the building design to reduce the size of the building compartments and so satisfy building regulations. The decision to include a firewall was consequently made once the primary structure was constructed. As the design and build contractor's project manager commented:

> We were literally putting up a building when we found that our areas were over what we thought they were. Whereas normally at the conceptual design [stage] you would see it and stop and evaluate it, whereas having been committed to a building out there, we had to make the decision [to include a fire wall].

The original plan was to erect the firewall using a tilt-up panel method of construction. However, penetrations would need to be made in the wall to accommodate plant and services and, at that stage, the dimensions and locations of penetrations were not known. As a result of this uncertainty it was decided to construct the wall using block work to allow for penetrations to be made more easily when the building's equipment and services design was finalized. The project manager commented: 'The equipment contractors were directly contracted to [the client] and they were hard to pin down… so this issue has see-sawed back and forth with the issues that we have had with the openings.'

The local fire authority also played an important role, as it became apparent that the building design deviated from the specification standards contained in the building regulations, necessitating approval of the firewall design by the fire authority. Notwithstanding a decision to construct the building using fire retardant panels, the fire authority advised that they would not support the original building design because the design did not provide full perimeter access for fire appliances.

Once the plant and equipment design was finalized, the design team discovered that the penetrations required in the firewall were considerably larger than the 600 mm^2 allowed for in the existing block work wall. This would necessitate re-work, and also compromise the fire integrity of the wall. Work commenced to enlarge the penetrations, presenting specific health and safety risks to workers involved in demolishing sections of the block work wall. Once the plant was installed, the installation contractor then advised that the openings in the block work wall could have been 40% smaller.

To maintain the integrity of the firewall, the penetrations were in-filled to the recalculated sizing. However, this reconstruction had to take place after the fixed plant was already installed and workers had restricted access to the work area. The construction of the penetrations required that the block work be cut and then flashed with stainless steel to adhere to the food safety regulator's requirements. While the openings were not high in the wall, scaffolding was required to provide access. The openings in the firewall remained a subject of contention. The fire authority maintained the block work wall could no longer act as a firewall when it included penetrations. In the opinion of the fire authority, the building was an oversized single building that required a sprinkler system to comply with the building regulations.

An assessment was commissioned from a fire engineer who advised that fire tunnels would be required either side of the wall to stop the spread of fire, smoke, and heat. The

---

Case Example 3.2 (Continued)

size (or length) of the tunnels was to be proportional to the size of the openings – the larger the opening, the longer the tunnel. However, limited space was available for constructing fire tunnels as fixed plant had already been installed either side of the firewall. The original design for the tunnel required a 2.5 m length, for which there was insufficient space. A reduction in the size of the openings permitted a reduction in tunnel length to 1.8 m. The construction of the fire tunnel commenced without the fire authority's approval in order not to fall behind the project schedule. In the event, the fire authority did not approve this design, insisting on installation of a full sprinkler system. To obtain approval for the building design, the client agreed to retrofit the building with a sprinkler system after the start-up of production.

The late inclusion of a sprinkler system into the design meant the installation presented specific safety challenges as workers needed to negotiate existing plant and services located in the ceiling, a confined space. Another area of safety concern was access to the underside of the ceiling to install sprinkler heads. Fixed plant and equipment had been installed in the building, which could not be moved to provide space for access equipment. Further, the production plant was operational when the sprinkler system was installed, providing only a short window of opportunity to carry out the work.

(*Source:* Lingard et al. 2013b)

Case Example 3.2 reveals the tensions and trade-offs that can arise when designing a facility for safe construction and operation. It also shows the role played by external stakeholders and the instability of design. Little or no guidance is provided in published practice guides on how to identify and manage conflicts and trade-offs between safety and other design criteria, or between safety in one or more stages of a facility's lifecycle. Yet, it is important that decision makers recognize and explicitly address these trade-offs when making design decisions.

3.6   The Case for Integration

3.6.1   Early and Effective Consideration of Safety in Design

Swuste et al. (2012) comment that the design phase of a construction project offers the greatest potential to positively influence safety. This argument is linked to Szymberski's (1997) concept that the ability to influence safety deteriorates rapidly as the project passes through the pre-construction stages. At the commencement of construction, the ability to influence safety is very low.

Hare et al. (2006) suggest health and safety can be more effectively integrated into early project design decision making by involving constructors in the project as early as possible, and creating opportunities for two-way communication between designers and people with construction knowledge. Recent research in the Australian construction industry supports these findings. This research formed part of an international benchmarking study of safety in design. Data were collected from a total of 23 construction projects – 10 in Australia and New Zealand and 13 in the USA. In each project, specific elements or components of the building (or other facility) were selected. The total

Table 3.2 The hierarchy of control.

Level 1	Eliminate a hazard altogether. Most effective because a hazard is removed physically from the work environment.
Level 2	Substitution of a hazard. Something that produces a hazard is replaced by something less hazardous.
Level 3	Engineering controls. People are physically isolated from hazards.
Level 4	Administrative controls. These include safe work procedures, or using job rotation to limit exposure to a hazard.
Level 5	Personal protective equipment. This is the least effective control because it is the least reliable.

number of elements in the analysis was 43. Elements included roof structures, sewerage systems, retaining walls, a pedestrian bridge, and foundation systems. Project stakeholders involved in planning, designing, and constructing the buildings (or other facilities) were interviewed. Interviews explored design decisions made for each element, the construction process for the element, and how health and safety hazards were controlled during construction. Interviews also explored the timing and sequence of key decisions about each element and the influences that were at play as design decisions were made. A total of 288 interviews were conducted (185 in Australia and 103 in the USA). The average number of interviews per element was 6.7. For each building (or facility) element, a score was generated that reflected the quality of health and safety risk controls implemented during construction. This score was based on the Hierarchy of Control (HOC).

The HOC is a widely accepted approach to controlling workplace risks or hazards (see, for example, Manuele 2006). The HOC classifies hazard control measures into five levels of effectiveness. Level 1 is the most effective method of control. Level 5 is the least effective method of control. Levels 1, 2, and 3 are technological risk controls. They involve changes to the physical work environment. Levels 4 and 5 are behavioural risk controls. They seek to alter how individuals and teams undertake their work (Table 3.2). It is often argued that safety in design will increase opportunities to implement higher order (technological) controls for health and safety risk (see, for example, Gangolells et al. 2010). However, until recently, there was little empirical evidence to support this claim.

In the Australian–USA research collaboration, design outcomes were scored according to the quality of risk control outcomes that were realized. Each HOC level was given a rating ranging from one (personal protective equipment) to five (elimination). The risk controls implemented for hazards presented by each element of the building or facility being considered were assigned a score on this five point scale. In the event that no risk controls were implemented, a value of zero was assigned. Using these values, the mean HOC score for each feature of work was generated. These scores are presented in Table 3.3 by country, project delivery method, and industry sector. Australian cases in the analysis had significantly higher average HOC scores than the USA cases, which may reflect the differences in legislative environments related to safety in design (see policy responses above).

The point in time was recorded at which a risk control solution was identified; that is, whether this occurred in the project's pre-construction or construction stage. For

Table 3.3 Mean HOC scores by country, project delivery method, and industry sector.

Case descriptor	Mean HOC score	Standard deviation
Country		
USA	2.48	0.311
Australia	3.69	0.671
Delivery method		
Collaborative	3.36	0.632
Accelerated	2.98	0.820
Design-bid-build	2.71	0.602
Design and build	3.38	0.233
Sector		
Heavy engineering	3.33	0.844
Residential	3.02	0.777
Commercial	2.72	0.649
Industrial	3.13	0.807

each building/facility element, the number of safety solutions selected during the pre-construction stage was expressed as a percentage of the total number of safety solutions for that element – the percentage reflected the extent to which safety was considered early in the project lifecycle.

Figure 3.3 shows the relationship between:

- the extent to which safety solutions were considered and decided upon before construction commenced (that is, in the planning or design stages of the project); and

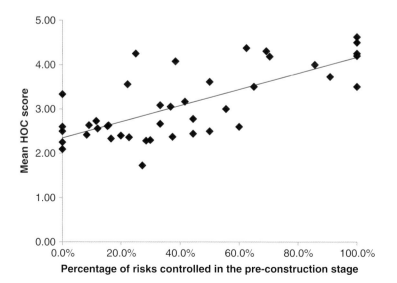

Figure 3.3 Relationship between pre-construction health and safety decision making and quality of risk control outcomes.

- the quality of risk control outcomes realized (that is, the average HOC score).

A positive relationship was found, meaning that the greater the proportion of safety solutions that were identified and chosen before construction commenced, the better the quality of the safety risk controls realized. This relationship was also statistically significant (Lingard et al. 2015a,b).

This research provide supporting evidence for the link between considering workers' health and safety early (in pre-construction stages of the project lifecycle) and implementing effective (technological) controls for health and safety risks.

3.6.2 Integrating Process and Product Knowledge

There are considerable benefits to involving constructors early in design decision making. Song et al. (2009) identified three primary benefits:

- constructors have specialized training, knowledge, and experience in applying construction materials and methods;
- constructors are in the best position to provide advice about health and safety hazards/risks and ways to mitigate them in construction activities; and
- because they are responsible for a project's construction operations, constructors have a strong motivation and interest in ensuring work is performed with minimal risk to health and safety.

The Australian–USA research also investigated whether involving constructors in design decision making produced better health and safety risk control outcomes.

To investigate this, a technique known as social network analysis was used. Social network analysis is an analytical tool that studies the exchange of information between people who make up a network. Social network analysis was used to map the social relations between project participants in each of the Australian case studies (elements of buildings or other facilities). The constructors' position of 'centrality' in the social networks was quantified. 'Centrality' refers to the extent to which a person is connected to other people – that is, the ratio of the number of relationships the person has relative to the maximum possible number of relationships they could have. Centrality is sometimes used as an indicator of the power or influence a person has within a network. In the case study projects, the constructors' centrality was measured during the design stage of the project. The relationships between members in a social network can be mapped to produce a 'sociogram', which is a graphic representation of the position and importance of participants within a network.

An example sociogram is shown in Figure 3.4. The sociogram shows three groups.

1) On the right-hand side of the network are demand-side stakeholders, including the owner, owner's engineer, and project manager.
2) On the left-hand side of the network are key supply-side stakeholders, including the concreters and steel erectors.
3) Also on the left-hand side of the network are stakeholders who supply design-related information and services to the network – the checking engineer and building surveyor.

The design and construction contractor had direct links with the majority of other network participants during design decision making in this project. As the central

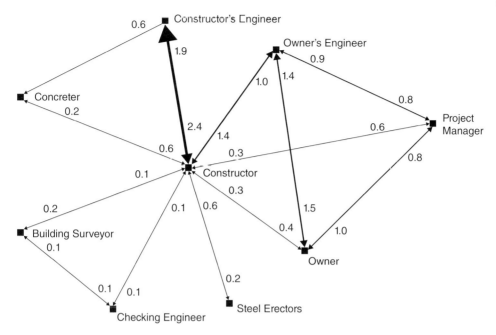

Figure 3.4 Social network for the steel column and roof design at a food processing and storage facility.

actor, the contractor connected the three groups and was able to use this central position in the network to identify and solve health and safety issues before construction commenced. The network pattern shows that the constructor took advantage of direct information ties with suppliers and subcontractors (steel erectors and concreters) to redesign various components to improve health and safety, while still meeting the

---

Case Example 3.3 Design and Construction of Steel Columns and Roof Structure at a Food Processing and Storage Facility

An initial concept design was developed on behalf of the client to accommodate operational requirements for the facility. The concept design included a steel-framed structure consisting of three spine trusses supported by five rows of steel columns. To maximize useable floor space, the columns were positioned in the middle of product stacks rather than at the ends of the rows.

The design and construction contractor suggested eliminating one row of columns. This design alternative required fewer columns to be lifted and manoeuvred into place, reducing health and safety risks associated with lifting operations. The contractor also suggested revising the roof design by using trussed rafters connecting to the main spine trusses, instead of using steel 'I beams' as rafters. Fabricating rafter trusses was slightly more expensive, but these trusses weighed less than I beams and could be manufactured offsite. The reduced weight of the roof enabled the use of smaller sections for supporting columns. It also made erecting and installing the roof quicker and easier.

(Continued)

Case Example 3.3 (Continued)

All supporting columns were fitted with a bearing plate allowing trusses to be supported temporarily while connections at each end were bolted. This reduced the need for propping and manual handling associated with installing and dismantling props. It also freed the area around the columns, and under the trusses, of any obstacles or trip hazards that props may have caused. At the same time, this design solution reduced the extent of work required at height to connect the trusses to the columns, and reduced the health and safety issues associated with suspended loads. As the client's engineer commented:

> [The constructor has] got quite a good, what I call a bearing type detail, so you can actually put the trusses up and have them take the gravity load away before you start trying to put the bolts in. And that's one of the major concerns [on another similar project] is that we should have picked it up when we did the structural check, but of course we just checked the structure rather than checking the buildability.

The structure was designed so that erection could be done in self-supporting sections. This allowed the builders to start at one end of the building and move progressively along the length of the building. This method enabled the constructor to ensure crane lifts were within safe reach tolerances, without having to extend the crane's arm over already constructed portions of the structure. To ensure the constructability of the facility before the start of construction work, the main constructor involved subcontractors in reviewing design and erection/installation sequences.

The resulting safety in design solutions resulted in a HOC score of 4.2.

---

owner's operational requirements and complying with relevant regulatory requirements (see Case Example 3.3).

All the project cases were statistically analysed to examine whether the construction contractors' network position in the design stage of a project was linked to health and safety outcomes.

The frequency with which communication flowed from the construction contractor to other parties during the design stage of the project was measured in each of the 13 project networks. Projects were divided into those which produced higher than average and lower than average health and safety performance outcomes (in terms of implementing upper HOC level versus lower HOC level risk controls). The results showed a statistically significant difference with better than average health and safety risk control outcomes in projects in which the construction contractor was in an influential position. That is, in projects where more upper level health and safety risk control measures were applied, the construction contractor was more engaged in the design stage in frequently providing information to other decision-makers. By contrast, in projects where the health and safety risk controls applied during the construction

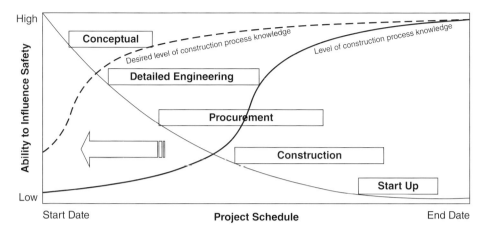

Figure 3.5 Time–process knowledge influence curve. *Source:* Lingard et al. (2015c).

stage of the project were less effective than average, the construction contractor played a less significant role in project communication in the design stage (Lingard et al. 2014).[1]

These results provide evidence to support claims previously made that when design decisions are informed by practical experience of construction processes, better health and safety outcomes are likely to be realized (Gambatese and Hinze 1999).

These results indicate that strategies to elicit constructors' process knowledge during the early stages of a construction project are likely to improve the effectiveness of safety in design activities, and facilitate adoption of technological/upper level (rather than behavioural/lower level) controls for health and safety risks. That is, there is a need to push construction process knowledge upstream to make it available to decision makers in the design stage of construction projects.

Figure 3.5 builds on the time/safety influence curve developed by Szmberski. The solid line illustrates the availability of construction process knowledge to decision makers as the project progresses from design through procurement to construction. As can be seen in the early project stages, the available process knowledge is limited; however, construction process knowledge availability increases as the project progresses, and increases dramatically at the procurement stage. The large arrow indicates the desired shift to provide greater depth and quality of construction process knowledge to decision makers earlier in the project lifecycle, as represented by the dashed line.

1 This work was supported by Cooperative Agreement Number U60 OH009761, under which RMIT is a subcontractor to Virginia Tech, from the USA's Centers for Disease Control and Prevention (CDC), National Institute for Occupational Safety and Health (NIOSH). Its contents are solely the responsibility of the authors and do not necessarily represent the official views of the CDC NIOSH.

3.7   Integrating Mechanisms

3.7.1   Collaborative Project Delivery Mechanisms

The extent to which detailed knowledge of construction processes is available to decision makers is influenced by the project delivery mechanism selected by the client. For instance, traditional project delivery methods are linear and sequential, and separate the design and construction functions. In contrast, there are other forms of project delivery that encourage more involvement of designers in addressing workers' safety, especially those forms in which design and construction work are undertaken by the same entity or those which create a partnership between the design and construction teams, closing the gap between these two parties. These more integrated forms of delivery can facilitate the use of construction knowledge at the design stage, and encourages designers to address construction issues (including health and safety hazards) in their decision making (Gambatese et al. 2005).

Recent analysis of performance data collected at a large rail infrastructure project in Australia reveals that packages of work delivered through more collaborative delivery mechanisms, such as alliances, tended to demonstrate better health and safety performance than packages of work procured using a more traditional design and construct delivery mechanism.

Although improved health and safety are often claimed to result from collaborative or integrated approaches to project delivery, some researchers caution that the implied link is not straightforward. Ankrah et al. (2009) observe that the procurement method will not generate, as a matter of course, a positive cultural orientation to health and safety. Similarly, Atkinson and Westall (2010) point out that an integrated project delivery approach is no guarantee of improved safety outcomes. Integrated project delivery mechanisms create favourable conditions for integrating health and safety into construction project planning and design activities, but actual health and safety improvements are likely to occur as a direct result of increased communication and information exchange among project participants and stakeholders.

3.7.2   Sharing Knowledge

Various models have been used to capture and make construction process knowledge available to design decision makers. Some have sought to codify this knowledge and make it available in the form of knowledge-based systems (see, for example, Robertson and Fox 2000; Cooke et al. 2008). Cooke et al. worked with a multi-stakeholder group of industry experts, comprising safety professionals, engineers, facilities managers, and construction personnel. Participants were asked about the different features of roof design, such as pitch, surface material, layout, and accessibility, and the safety implications of alternative design options for each feature were explored. The knowledge was used to create multiple interactive weighted decision trees that represented all the different options in combination and produced a determination, based on the experts' evidence, about the extent to which a particular roof design would present high, medium, or low risk of a worker falling from height. This knowledge was embedded into a prototype online decision support tool, known as ToolSHeD (Tool for Safety and Health in Design). ToolSHeD could be used by designers to assess their design, and to review and change their design as needed to produce the desired level of risk. The ToolSHeD concept was well received by design professionals as the information it contained was presented in a way that could be easily understood and applied to practical design decisions. However, the knowledge base underpinning it was large and cumbersome, and there were unresolved challenges about how such a knowledge base could be kept up to date to reflect new technologies and safety in design solutions. The ToolSHeD experiment raised interesting questions about how knowledge of work processes and health and safety could be made available to designers to enable improved design decision making in the interest of workers' health and safety.

3.7.3   Infographics and Visual Communication

There is increasing use of visual methods to capture and communicate complicated scientific or technical information (Brumberger 2007a; Estrada and Davis 2015), and there is a growing understanding that 'the visual' provides a powerful means of representation and argumentation (Pauwels 2000). Comai (2015) describes how well-designed visuals are not simple checklists of what to do next; they also provide suggestions and new insights that generate intelligent decision making. The need to pay more attention to visual thinking in the design and practice of communication is well recognized (Portewig 2004).

In the field of architectural design, Whyte et al. (2007) also explore ways that visual practices and objects are used to facilitate iterative design development and collaborative decision making. There is evidence that using images to convey meaning evokes different types of knowledge, as compared with using the written or spoken word (Harper 2002). Research also shows the effectiveness of images for communicating mechanical and spatial relationships in ways that are hard to capture with words alone (Houts et al. 2006).

Infographics are a particular type of visual communication tool, increasingly used to communicate information in many fields. Infographics – an abbreviation for informational graphics – are defined as graphic representations of information (Lancow et al. 2012). Infographics are now widely used in the mass media and can take many forms (Lester 2011). For example, they range from basic arrangements of facts and figures to annotated charts, cartoons, maps, and complicated interactive graphics. They can also be static, animated, or interactive (Otten et al. 2015). Whatever their form, infographics are not just forms of artistic expression. Infographics play a key role in telling a story and should not be seen as secondary to text (Lazard and Atkinson 2014).

Infographics have recently been trialled as a method for capturing and communicating work health and safety knowledge for construction design professionals. Figure 3.6 shows an example infographic relating to work health and safety aspects of a façade design.

When infographics were provided to design professionals in a workshop format they were found to increase designers' ability to recognize health and safety hazards. Brumberger (2007b) describes visual thinking as an active problem-solving process in

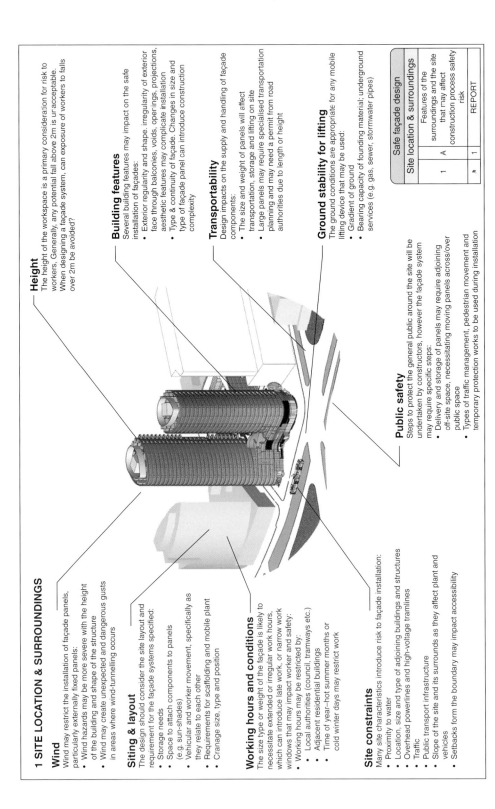

1 SITE LOCATION & SURROUNDINGS

Wind

Wind may restrict the installation of façade panels, particularly externally fixed panels:

- Wind hazards may be more severe with the height of the building and shape of the structure
- Wind may create unexpected and dangerous gusts in areas where wind-tunnelling occurs

Siting & layout

The design should consider the site layout and requirement for the façade systems specified:

- Storage needs
- Space to attach components to panels (e.g. sun-shades)
- Vehicular and worker movement, specifically as they relate to each other
- Requirements for scaffolding and mobile plant
- Cranage size, type and position

Working hours and conditions

The size type or weight of the façade is likely to necessitate extended or irregular work hours, which can introduce late work, or narrow work windows that may impact worker and safety:

- Working hours may be restricted by:
  - Local authorities (council, tramways etc.)
  - Adjacent residential buildings
  - Time of year–hot summer months or cold winter days may restrict work

Site constraints

Many site characteristics introduce risk to façade installation:

- Proximity to water
- Location, size and type of adjoining buildings and structures
- Overhead powerlines and high-voltage tramlines
- Traffic
- Public transport infrastructure
- Slope of the site and its surrounds as they affect plant and vehicles
- Setbacks form the boundary may impact accessibility

Height

The height of the workspace is a primary consideration for risk to workers. Generally, any potential fall above 2m is unacceptable. When designing a façade system, can exposure of workers to falls over 2m be avoided?

Building features

Several building features may impact on the safe installation of façades:

- Exterior regularity and shape: irregularity of exterior face through balconies, voids, openings, projections, aesthetic features may complicate installation
- Type & continuity of façade. Changes in size and type of façade panel can introduce construction complexity

Transportability

Design impacts on the supply and handling of façade components:

- The size and weight of panels will affect transportation, storage and lifting on site
- Large panels may require specialised transportation planning and may need a permit from road authorities due to length or height

Ground stability for lifting

The ground conditions are appropriate for any mobile lifting device that may be used:

- Gradient of ground
- Bearing capacity of founding material; underground services (e.g. gas, sewer, stormwater pipes)

Public safety

Steps to protect the general public around the site will be undertaken by constructors, however the façade system may require specific steps:

- Delivery and storage of panels may require adjoining off-site space, necessitating moving panels across/over public space
- Types of traffic management, pedestrian movement and temporary protection works to be used during installation

Safe façade design		
Site location & surroundings		
1	A	Features of the surroundings and the site that may affect construction process safety risk
2	1	REPORT

Figure 3.6 Infographic showing health and safety aspects of the site environment. *Source:* reproduced from Lingard et al. (2018).

which familiar objects and processes are seen in new ways from different perspectives. The results reveal that after workshop participants viewed the infographics they demonstrated deeper thinking about construction workers' health and safety and the implications of design decisions. Before participants viewed the infographics, they were able to identify physical health and safety hazards related to issues in the immediate work environment; for example, hazards associated with falling from height or being struck by a moving load. These hazards were relatively easy for participants to envisage in the example façade design scenario provided to them in the workshop. However, after participants viewed the infographics, they were able to identify many more design-related issues that could potentially create a situation in which the risk of injury or harm was increased – that is, shaping factors. These included issues relating to component quality and supply chain issues, working schedule arrangements, and erection sequencing. Ergonomic/manual handling hazards were also identified after participants had viewed the infographics.

The designers commented that the infographics enabled them to consider aspects of a design in a more holistic way to better understand the interconnectedness of the various design elements. One participant noted: 'I suppose at a glance you can see the whole environment. Whereas when something's in writing you just focus on the one issue and not the whole environment. It's a much more global thing.' Other participants described how the infographics reinforced their existing knowledge and 'brought to the fore the risks and got you to look a bit deeper into a situation'. The potential for infographics to improve collaboration and create a shared understanding of workers' health and safety was also noted: 'because people do have different backgrounds, different ways of looking at things.'

The benefits that flow from using visual communication are likely to be enhanced when multiple stakeholders contribute to visual representations of health and safety information. We return to this theme in Chapter 5 when we consider the importance of working collaboratively and developing shared mental models of working safely.

3.8 Conclusions

In this chapter we have presented the case for considering construction workers' health and safety during the early stages in a project lifecycle. In particular, we have examined the potential benefits associated with considering WHS when key decisions are being made in the design stage of a project. However, we have also revealed how challenging the integration of WHS can be, given the sociotechnological complexity of design work in the construction industry. In this context decisions can be made by parties who are distant from the construction work, and who may have little or no awareness of the implications of their decisions. Further, decisions made at one point in time can have a cumulative impact as they impact subsequent decisions made by others involved in design work.

We also highlight the potential benefits to be gained from ensuring that design decision making is informed by an understanding of construction methods, materials, and technologies. Thus, it is important to consider the WHS implications of the design of the product to be constructed, as well as the construction process. Our research has shown that when the technical and experiential knowledge of people who understand the processes of construction is accessed and used to inform product design decisions, more effective WHS risk controls are ultimately realized.

Yet, achieving this outcome is easier said than done in a fragmented design environment and those who possess in-depth process knowledge may not be involved when important design decisions are being made. Specialist subcontractors, for example, may only be engaged once a principal contractor has been appointed and much of the design work is complete. Visual approaches to capturing and conveying construction process knowledge to design decision makers have proven effective. The visual power afforded by building information modelling technologies the use of virtual prototyping and serious games provides considerable potential for capturing and transferring WHS information and helping to bridge the knowledge gap that currently exists between those with responsibility for designing the construction industry's products and processes.

Discussion and Review Questions

1 What are the structural impediments to integrating work health and safety considerations into design decision making in construction projects?

2 Who is the designer in construction projects? Should responsibility for safety in design outcomes be allocated to individuals or collectively shared?

3 How can work health and safety outcomes be improved in the design of the construction industry's products and processes?

4 By what measures should the effectiveness of safety in design activities be assessed?

Acknowledgement

Research presented in this chapter was funded under different grants programmes.

The social network analysis and analysis of early project decision making and WHS outcomes was supported by Cooperative Agreement Number U60 OH009761, under which RMIT was a subcontractor to Virginia Tech, from the USA's Centers for Disease Control and Prevention (CDC), National Institute for Occupational Safety and Health (NIOSH).

The infographics research was funded under an Australian Research Council Linkage Project Grant (LP120100587). The infographic depicted in Figure 3.6 was conceived and developed by Prof. Nick Blismas.

4

Construction Workers' Health

Helen Lingard and Michelle Turner

School of Property, Construction and Project Management, RMIT University, Melbourne, Victoria, Australia

4.1   Introduction

4.1.1   A Neglected Issue

Managing safety risks in the construction industry has spurred much focus and generated great deal of effort to reduce incidents and injuries. Managing occupational health risks has attracted far less attention (Constructing Better Health 2018). The relative neglect of workers' health may be explained by the long latency periods of many illnesses, difficulty disentangling work-related factors and other factors contributing to poor health, and the fact that the link between exposure and health outcomes is sometimes difficult to understand. The time lag between exposure to health risks and illness can create complacency in an industry such as construction because the project-based work, and transient (often casualized) workforce, can make it difficult to trace an illness back to exposure during a particular employment episode. It has also been argued that standard management approaches implemented for managing safety risks are insufficient to produce effective risk controls for work-related illness (Sherratt 2015). In Australia, construction workers report they are frequently exposed to workplace health risks, including airborne hazards, vibration, chemicals, and biomechanical hazards. In many instances, no risk control measures are implemented, or there is a heavy reliance on workers' use of personal protective equipment (PPE) to protect themselves from harm (Safe Work Australia 2015b).

Attention paid to construction workers' health has grown in recent years, bolstered by a growing awareness of the costs of ill-health in terms of sickness absence, reduced worktime, and diminished productivity. However, in her review of the health of Britain's working age population, Dame Carol Black (Black 2008) found employers had limited understanding of the evidence base that supports the business case for investment in health and wellbeing. Black also found an institutionalized view that it is inappropriate for people to be at work unless they are 100% fit. Employers have insufficient processes for helping workers remain in the workforce or return to work following illness. A growing awareness of the social, as well as economic, impacts of poor health in people of working age has focused the attention of government agencies, large employers, and clients in the construction industry.

Integrating Work Health and Safety into Construction Project Management, First Edition.
Helen Lingard and Ron Wakefield.
© 2019 John Wiley & Sons Ltd. Published 2019 by John Wiley & Sons Ltd.

This attention is overdue because the magnitude of the occupational health problem is substantial. In the UK 1.4 million workers suffered from work-related ill health (new or long-standing) in 2017/18. Many of these workers experience stress, anxiety or depression (44%), or musculoskeletal disorders (35%). Further, 12,000 lung disease deaths occur in the UK each year are that are linked to past exposures at work (Health and Safety Executive 2018b). Also, in the UK, construction workers are at least 100 times more likely to die from a disease caused or made worse by their work as they are from a work-related injury (Institute of Occupational Safety and Health 2015). This is consistent with figures in Australia, where it is estimated that 250 Australian workers die from an injury sustained at work each year, yet over 2,000 workers die from a work-related illness each year (Safe Work Australia 2012). In Australia, mental illness costs businesses A\$10.9 billion per year (BeyondBlue 2014). Mental illness in the construction industry is 5% higher than the Australian average, with one quarter of construction workers experiencing mental illness. The high incidence of mental illness triggered a call to focus on preventative programmes in the workplace to improve workers' mental health (WorkSafe Victoria 2016). Startling statistics also relate to the occurrence of suicide in the construction industry. Mates in Construction (MIC) estimate Australian construction workers are six times more likely to die as a result of suicide than as a result of a safety incident at work.

4.1.2 An Integrated Approach to Managing Workers' Health

There has been growing support for an integrated approach to prevent injury and to advance health and wellbeing in the workforce (Anger et al. 2015; Pronk 2013; Sorensen et al. 2011). In recognition of this, the USA launched the Total Worker Health® programme in 2011, led by the National Institute for Occupational Safety and Health (Schill and Chosewood 2013; Sorensen et al. 2011). The move towards an integrated model of worker health recognizes that preventive occupational health programmes seek to manage specific components of workers' health which arise due to occupational health hazards. The health hazards workers are exposed to are linked intrinsically to the environment in which they occur. External factors, such as organizational and project outcomes and the way in which work is organized, interact to impact on health behaviour and health outcomes. Another key factor contributing to health is individual choice relating to personal lifestyle factors. While individual choice is important in shaping health and wellbeing, the choices individuals make are also subject to social, cultural, and environmental influences. Research has indicated that organizations which implement health promotion and disease prevention programmes must consider the broader environment, in addition to individual factors, and identify how organizational characteristics contribute to poor health (Ettner and Grzywacz 2001; McLeroy et al. 1988; Lingard and Turner 2015).

This chapter is arranged according to the key factors informing the integrated model of worker health. Developed through industry-based research, the model appears linear and static in nature. However, its key components are fundamentally dynamic and interdependent. The model is presented in Figure 4.1. With reference to this model, the chapter describes some of the most significant occupational health risks in construction work and emphasizes the need to implement effective controls for known hazards. The importance is identified of designing for construction workers' health as a means

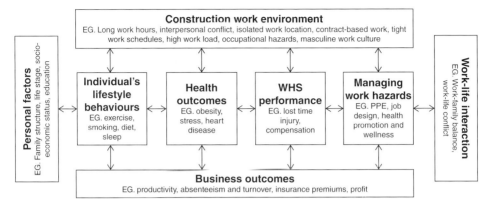

Figure 4.1 Integrated model of worker health.

of improving the quality of risk control currently implemented for work-related health hazards.

This chapter considers the relationship between occupational health hazards and risks of so-called 'lifestyle' diseases, often the focal point of investment in workforce wellness programmes. Such wellness programmes are critically appraised, drawing on evidence from construction industry case studies. The case studies highlight the importance of factors relating to the design and organization of work in addressing workers' health. The need for a multi-level systems approach is emphasized.

Finally, the case is made for a more integrated approach to managing workers' health and safety. Critical to this is the application of an evidence-informed participatory approach to designing healthy work processes and workplaces.

The case study below describes a construction industry client's integrated approach to supporting worker health on at an industrial facility construction project. In particular, the client modified conditions of the *construction work environment* and *managed work hazards*. This acted to positively impact on *work health and safety (WHS)* performance and *work–life interaction*. The case study suggests this multi-level approach supported worker health and contributed to positive *business outcomes*. Importantly, the organization took a long-term approach to worker health by incorporating it into the project's aim.

Case Example 4.1 Client Drives Healthy Working in New Zealand Construction Industry

The client, an organization undertaking a NZ$73 million factory expansion, made workers' health and safety its first priority. This is unlike most construction projects, which are largely driven by time and budget. A senior manager explained the client's approach to the project:

> We wanted to create a legacy that health and safety are the most important things. We had a slogan that emerged that we pushed throughout the length of the project which was safety one, quality second and timeline third. But I think we didn't

(Continued)

Case Example 4.1 (Continued)

just want to say these things, we wanted to actually act on them. We wanted our contractors to regain some type of life balance. There has been a history where contractors have worked hard, big hours. We wanted to give them the opportunity to regain life balance which would have a positive impact on how they approached the project. We wanted to also strive towards creating a positive safety culture. We wanted to have a very clear picture of what was acceptable and what was not acceptable.

The senior management team implemented a programme to support the 'safety first, quality second, timeline third' approach that was applied during the 18-month construction project. The programme's intent was to create an integrated approach to wellness for workers, and the key components of the programme were leadership, culture, communication, policy, and practice. The programme was supported at all levels of the organization, from board to supervisors. This was a key feature of the programme and one which had a critical impact on project success. The client also acknowledged the key role of leaders in the performance of their teams, and so invested in a leadership development programme.

Communication was carefully considered throughout the project's duration. Messaging was kept simple and repetitive, and a variety of methods were used. For example, senior managers regularly went onsite and had safety conversations with workers. During high stress times, such as commissioning, strategies were implemented to create opportunities for communication, such as daily meetings and barbeque lunches.

A Fatigue Management Policy (FMP) was implemented at the site, which stated that working time should not exceed 60 hours per week. The policy was incorporated into the procurement strategy. At the front-end of contract management, contractors were made aware of expectations of hours due to the wellness focus at the project. Given the FMP's emphasis, contractors were invited to negotiate timelines with the client to ensure safety and quality standards were maintained. A senior manager explained:

> Rather than dictating to them a date that it had to be completed by, we negotiated with them as to how many people we can put on the work front and how that worked out to a timeline.

If more than 60 hours were planned, the client asked contractors to advise in advance how they would manage that week. The client challenged contractors to consider extra resources to support the tail-end of such weeks, and identify which activities could be brought to the front of the week so that only low-risk work was undertaken at the end of the week.

While a key aim of the policy was to prevent fatigue, it was also acknowledged that enforcing the policy enabled workers to spend time with their families. A worker told one of the senior managers: 'I've done 55 hours this week. I'm going home, and I'm going to spend the weekend with my family. I haven't done that before on a project.'

Project culture was an area which the client proactively developed and managed. An open and considered approach to communication, problem solving, and conflict

Case Example 4.1 (Continued)

management was applied at the project. The client had purposively moved away from the 'policeman-type role' and taken on a consultative and educative approach. A senior manager commented:

> Lecturing people and hauling them over the coals over an incident isn't progressing the health and safety space for the project... but being proactive and working with them and finding better ways of doing things is.

There was an emphasis on creating trust with contractors: 'When we did have incidents, we never were out to hang anyone, once again trying to keep that trust thing going.'

There were various benefits of the programme for the client, and some of these are summed up by a senior manager who reflected on what usually happens on a project, and how this project differed:

> Usually, the last two weeks before start up, people are running in all directions and there's rubbish all over the floor, there's electrical cables, you know there's people stressed. At this project it looked like everything was calm and in control ... people were having good quality conversations about how to install things, about how to wire something up, about how to weld. So you know that people are not fatigued. People have got time to think quietly about the best approach in how to install something. You know I saw a quality of workmanship and I think that we've spent $75 million dollars on this project. I don't believe we've replaced one valve, pipe, or instrument in $75 million dollars, which is extraordinary. We were well resourced as well, but a good part of that is that we look after people and when people are looked after they can think properly and think clearly and make good quality decisions.

Many contractors reported it was the best project they had ever worked on due to the focus on health and safety. A senior manager commented: 'We've had people coming up to us and saying this is the best project that they've ever worked on and they've been in the industry for fifteen, twenty years.'

The client aimed to leave a legacy that workers' health and safety is the most important factor on a project. The client challenged contractors to rethink their approach to construction-related activity, with a clear message that timeline should not drive activity. A senior manager reflected: 'I think people will be talking about this project for a while. There's already people talking about changes to what they do now.'

4.2 Work and Health

The World Health Organization (WHO) defines health as 'a state of complete physical, mental and social well-being and not merely the absence of disease or infirmity'. A review by Waddell and Burton (2006) found that work is generally good for both physical and mental health and wellbeing, and the benefits of work on health outweigh the risks of work on long-term sickness absence or work disability. Families without a working member are also likely to experience poverty and social disadvantage, and children in such families are reported to experience a higher incidence of psychiatric disorders and persistent health problems (Black 2008). Further, these disadvantages are likely to perpetuate poor health in future generations as social and health inequalities are not independent of one another. For example, poor childhood conditions, low levels of education, and blue collar employment are consistently linked to unhealthy behaviours and a propensity for health complaints in adult males (Lynch et al. 1997).

However, the impact of work on health depends upon the quality of jobs and employment (World Health Organization 2008a). In the report *Closing the gap in a generation*, the WHO and the Commission on the Social Determinants of Health identify fair employment and decent working conditions as critical in the attainment of health. When work is fair and conditions are good, the experience of work is likely to be positive and contribute to good health; for example, by providing financial security, social status, personal development, improved social relations, self-esteem, and protection from physical and psychosocial hazards. However, adverse work conditions have a negative impact on workers' health; for example, exposures to physical and psychosocial health hazards, precarious employment, job insecurity, and stressful working conditions are all linked to poor health.

The relationship between health and work is complicated and reciprocal as being in good health is also a determinant of sustained workforce participation (Australian Institute of Health and Welfare 2010). Research shows that poor health often precedes early retirement (van den Berg et al. 2010; de Wind et al. 2013). Construction work is both physically and psychologically demanding. Exposure to psychosocial health risks and mental health symptoms are related to physical health (Abbe et al. 2011). Borsting Jacobsen et al. (2013) report mental distress in construction workers is strongly significantly associated with the experience of lower back pain, having two or more pain sites, and the experience of injury. Boschman et al. (2014) also found mental health problems significantly impact on the physical ability of bricklayers and construction supervisors to perform their work.

Consequently, before they reach retirement age, construction workers in many countries are reported to suffer from permanent work incapacity or forced to stop working due to health problems (Brenner and Ahern 2000; Welch 2009; Oude Hengel et al. 2012). Even compared to other blue collar occupations, construction workers experience high levels of work incapacity, and longitudinal cohort studies reveal that up to two-thirds of construction workers stop work as a result (Arndt et al. 2005; Siebert et al. 2001). In Australia, the Construction, Forestry, Maritime, Mining and Energy Union has expressed concern about calls to increase the statutory pension age due to the likely impact on workers engaged in physically demanding construction work (Collett 2014).

To compound the problem, many industrialized countries also face ageing populations. The increasing ratio of retirees to people in employment is placing strain on national and social welfare systems, and governments are actively encouraging people to work later into their lives. This makes maintaining good health even more important so that people can continue working as they age (Australian Institute of Health and Welfare 2010). As populations age, the construction industry in many countries is facing a labour shortfall. The median age of construction workers in the USA increased from 37.9 years in 2000 to 42.6 years in 2017; but between 2016 and 2026, the number of jobs for construction labourers in the USA will rise by 12.4% and jobs for construction managers will rise by 11.1%, compared to an average of just 7.4% across all other job categories (Bureau of Labor Statistics 2018). Given these demographic trends, helping construction workers to maintain good health and remain in employment is economically, as well as socially, important (Noone 2013).

4.3 Organizational Issues and the Design of Work

Recognizing that decent work is fundamental to workers' health and wellbeing, the International Labour Organization (ILO) has established an international agenda with four strategic objectives, as follows:

- promoting jobs;
- guaranteeing rights at work;
- extending social protection; and
- promoting social dialogue.

Decent work is underpinned by the expectation that workers are provided with safe and healthy conditions while at work. However, decent work also provides workers with sufficient time away from work to rest and recover from work demands, and to actively participate in family, leisure, and other social activities.

Work in the construction industry is characterized by exposure to significant physical and psychosocial hazards that have adverse health impacts. In the UK, Stocks et al. (2010) analysed instances of medically reported work-related ill-health among construction workers and found elevated rates of contact dermatitis, all types of skin neoplasma, non-malignant pleural disease, mesothelioma, lung cancer, pneumoconiosis, and musculoskeletal disorders (MSDs). Construction workers are exposed to many hazardous physical working conditions, including manual handling and exposure to vibration, noise, chemicals, and dust (Snashall 2005; Stocks et al. 2011). Stocks et al. (2011) report that incidence ratios for work-related illness differ between construction trades, with a higher risk of developing:

- long latency respiratory diseases among pipe fitters, electrical workers, plumbing and heating engineers, carpenters and joiners, scaffolders, and labourers in building and woodworking trades; and
- skin neoplasma among roofers, painters and decorators, and labourers.

Employment conditions in the construction industry have been linked to poor health because the effects of illness can be compounded by casual employment and limited access to sick leave, making recovery from injury and illness harder. For many workers, particularly those in manual or non-managerial work, income is dependent upon time spent at work. Consequently, financial pressures to return to work after injury or illness, before a full recovery has been made, can create more serious health problems in the longer term (Meerding et al. 2005).

Employers and other duty holders must, under WHS legislation, identify and manage occupational health hazards so that the risk of harm is reduced to as low as is reasonably practicable (ALARP). The following section describes some of these hazards and their health impacts.

4.4 Workplace Risk Factors

4.4.1 Musculoskeletal Disorders

The incidence of musculoskeletal disorders (MSDs) among construction workers is disproportionately high and contributes significantly to work disability (Inyang et al. 2012). Between 2009 and 2013, acute and chronic MSDs accounted for more than half (54%) of all workers' compensation claims in Australia (Safe Work Australia 2015a).

MSDs mainly occur when physical workload exceeds the physical capacity of the human body. Sometimes MSD occurs in a single event. In other cases it is the result of repeated trauma. Risk factors commonly associated with work-related MSDs in construction workers are repetition, force, awkward posture, vibration, and contact stress.

Specific tasks in construction work are associated with an elevated risk of MSD. For example, particular risk factors in bricklaying are working with a bent back, carrying and lifting, working with arms above shoulder height, and kneeling and stooping (Boschman et al. 2012). Parida and Ray (2012) also describe how manual material handling in construction work can contribute to MSD through tasks that involve poor and awkward postures, repetitive movements, hand–arm vibration, heavy lifting and handling, high physical stress, and overexertion.

Despite their prevalence, MSD risks in construction are often not well managed. There is a need to better understand them and consider ergonomic solutions to reduce the risks (Albers et al. 2005; Dale et al. 2016). The issue of work-related MSDs is discussed in greater detail in Chapter 8.

4.4.2 Noise

Construction workers are exposed to hazardous levels of noise that frequently exceed daily noise exposure standards (Leensen et al. 2011). Use of hearing protection devices by construction workers also tends to be poor, partly because of perceived difficulties in hearing and understanding speech communication and warning signals (Suter 2002). Noise-induced hearing loss (NIHL) is one of the most common occupational diseases among construction workers around the world (Arndt et al. 1996; Hong 2005; Kurmis and Apps 2007). Dement et al. (2005) report that 60.3% of construction workers examined in a USA medical surveillance programme were found to have material impairment of hearing. The incidence of hearing loss varied between trades, ranging from 47% among insulators to 78% among plumbers and steamfitters. Ringen et al. (2014) drew on the same 16-year medical surveillance programme to assess the risk of hearing loss over a working lifetime. For all construction trades combined, the lifetime probability of suffering hearing impairment was 73.8%, compared to 43.5% in a comparison group of administrative, scientific, and security workers and 53.1% in a control group of low-noise industrial workers.

4.4.3 Chemicals

In various construction activities, exposures have been documented to carcinogenic chemicals such as polycyclic aromatic hydrocarbons (PAHs), hexavalent chromium, diesel exhaust, and radon (Järvholm 2006). For example, PAHs are present in coal tar which was sometimes added to asphalt and is associated with an elevated risk of lung cancer. Like asbestos, the use of tar has been banned in some countries due to the serious health risks it presents. Epoxy resins are widely used in the construction industry and frequently cause allergic contact dermatitis. Workers who are sensitized and acquire an allergy to these products have an increasingly strong reaction every time they come into contact with them. It is estimated that one in five Dutch construction workers who use epoxy resins will develop an allergy to them (Spee et al. 2006). Allergic contact dermatitis is also commonly experienced by construction workers who become sensitized to cement (Lazzarini et al. 2012). Solvents in paints and glues are known to cause intoxication, liver damage,and nerve damage. Although solvent-based paints have been largely replaced by water-based paints, a study by Kaukiainen et al. (2005) found that painters have a high risk of respiratory symptoms and chronic bronchitis when compared to a control group of carpenters. Zorba et al. (2013) report high levels of skin complaints, including chronic and acute contact dermatitis and contact urticaria, compared with other occupational groups. Bitumen-laying workers are reported to suffer higher rates of acne than other occupations, probably due to exposure to chlorinated hydrocarbons (Zorba et al. 2013).

4.4.4 Airborne Hazards

Asbestos was previously widely used in construction. Asbestos is known to cause mesothelioma of the pleura or peritoneum (Welch et al. 1991). While it is no longer used in the manufacture of construction products, the long latency periods for asbestos-related disease means a continually rising incidence of mesothelioma among workers who were exposed. Previously widespread asbestos use has left a terrible legacy and exposure can still occur during the retrofitting or demolition of buildings.

Even materials that appear to be harmless can become dangerous when they are broken down (Spee et al. 2006). One particularly insidious occupational health issue affecting many construction workers is exposure to respirable crystalline silica (silica). Silica is present in commonly used construction materials, including sand, stone, concrete, and mortar. Silica is also used in the manufacture of many building products, including composite stone, bricks, tiles, and some plastics. When products containing silica are cut, crushed, drilled, polished, sawn, or ground, respirable dust particles are produced. Without proper protection, workers exposed to silica dust can experience serious health effects, including:

- chronic bronchitis,
- emphysema,
- acute, accelerated, or chronic silicosis
- lung cancer
- kidney damage, or
- scleroderma (a disease of the connective tissue of the body resulting in the formation of scar tissue in the skin, joints, and other organs of the body) (Safe Work Australia 2018a).

Exposure to respirable crystalline silica has been identified as the worst occupational lung disease crisis since asbestos (Atkin 2018). Silicosis is an incurable (often fatal) lung disease caused by breathing dust containing tiny particles of crystalline silica. While acute silicosis can develop after a short exposure to very high levels of silica dust, accelerated silicosis develops after exposures of 3–10 years to moderate to high levels of silica dust. Exposure to respirable crystalline silica associated with cutting engineered stone to construct kitchen and bathroom benches has been identified by occupational health experts as being a significant problem in the building construction industry, although many other construction tasks can expose workers to respirable silica unless the risk is properly managed.

Silica is a worldwide occupational health problem affecting construction workers. It is estimated that, in the European Union, 7000 cases of lung cancer attributable to silica exposure occur each year (Institution of Occupational Safety and Health 2015). The Institution of Occupational Safety and Health (2015) also reports that around 5 million people in the European Union are exposed to silica dust at work, with most of these workers (81%) employed in construction or in the manufacture of products used in construction (10%).

Many countries have established workplace exposure standards for respirable crystalline silica. In Australia, for example, the level of exposure that must not be exceeded is $0.1\,mg/m^3$ (i.e. 0.1 milligrams of silica per cubic metre of air) over an 8-hour time-weighted average (Cancer Council Australia 2017). This means that the maximum average airborne concentration of respirable crystalline silica when calculated over an 8-hour working day for a 5-day working week should not exceed this amount. Different countries have adopted different exposure standards for crystalline silica. Thus, in the USA the Occupational Safety and Health Administration (OSHA) Standard, which was implemented in September 2017, establishes an even lower exposure limit of $50\,\mu g/m^3$ (micrograms of silica per cubic meter of air), which equates to $0.05\,mg/m^3$ over an 8-hour work day (Occupational Health and Safety Administration 2018).

Despite the expression of these exposure standards as maximum concentrations over an 8-hour working day, it is important to appreciate that some work tasks have very high exposure 'peaks' even though they may not last long (Institution of Occupational Safety and Health 2015). Where workers have a working day longer than 8 hours or work more than 40 hours a week, employers must determine whether the time weighted average exposure standard needs to be adjusted to compensate for greater exposure during the longer work shifts, as well as decreased recovery time between shifts (Safe Work Australia 2018b). The expression of exposure standards as a time weighted average value also presents challenges for exposure monitoring and application of standards in construction as workers can often work shifts longer than 8 hours and the average weekly work hours of site-based workers typically exceed 40 (Lingard and Francis 2004). By their nature, construction project environments are also constantly changing, which could potentially impact on the reliability and effectiveness of monitoring.

Given the prevalence and seriousness of silica-related lung disease, it is critical that the construction industry reduce workers' exposure to respirable crystalline silica dust. The most effective form of risk control for respirable crystalline silica is to eliminate silica dust from the work environment. However, due to the many construction tasks that can give rise to respirable crystalline silica, this is not always possible. In some instances hazardous materials can be replaced with materials that are less hazardous. Work processes can also be changed to reduce the risk of exposure to respirable dusts, for example, using wet processes instead of dry ones (Workplace Health and Safety

Queensland 2013). Engineering controls, such as containment, ventilation, and suppression systems, can also reduce workers' exposures to respirable silica (Workplace Health and Safety Queensland 2013).

Depending on work tasks, workers should be provided with suitable respiratory protective equipment. Not all respiratory protective equipment provides sufficient protection and the risks inherent in particular tasks need to be carefully assessed so that the appropriate category of protective equipment can be selected. For example, Cole (2016) recommends the use of full-face P3 Powered Air Purifying Respirators for shotcreters and other workers who are located within a shotcreting exclusion zone or other high exposure areas. It is also very important that workers are trained in how to use respiratory protective equipment properly. Facial hair and stubble can significantly reduce the effectiveness of respiratory protective equipment, even if the correct type is provided (Frost and Harding 2015).

It is recommended that workers potentially exposed to respirable silica be regularly tested to ensure that their health is not being impaired, For example, Crossrail Limited in the UK introduced a health surveillance programme which included lung test functions every two months for workers exposed to high levels of dust (Crossrail Limited 2017). Results are shared with workers' managers and workers are referred to a lung respiratory specialist or a general practitioner if the lung function test shows impairment.

4.4.5 Emerging Hazards

Nanomaterials are increasingly used in the construction industry. These materials can improve the strength, durability, and performance of construction materials. For example, nanomaterials can improve heat insulation and provide self-cleaning and antifogging properties. Workers are already working with nanomaterials, yet there is a paucity of scientific research about the health effects of exposure to nanoparticles. Some research has linked exposure to nanoparticles to oxidative stress, fibrosis, cardiovascular effects, cytotoxicity, and possibly carcinogenicity (van Broekhuizen et al. 2011). However, complexity and uncertainty make it extremely difficult to apply existing risk management principles to nanotechnology (Marchant et al. 2008).

Lee et al. (2010) identify growing health and environmental concerns associated with using nanomaterials in the construction industry. They recommend that lifecycle exposure assessments be made for nanomaterials to understand the health and environmental risks associated with their manufacture, use, and disposal. Research could inform the design and development of nanomaterials that maintain performance but pose a reduced health risk. Where necessary, engineering controls (such as ventilation systems and dust collectors) and suitable protective equipment may need to be used during manufacture and use of products containing nanomaterials. Lee et al. (2010) also recommend personal monitoring and surveillance of workers' dermal, respiratory, and optical exposure. Particular attention may need to be paid to removing products containing nanoparticles during demolition, in much the same way as removing asbestos is now undertaken by specialist teams under carefully controlled conditions. Given increasing use of nanomaterials in the construction industry, research is much needed to ensure any associated risks to occupational or public health are properly managed (Breggin and Carothers 2006).

4.4.6 Psychosocial Hazards

Construction work is characterized by high demands and low levels of control. The pace of construction work is often driven by tight schedules, with financial penalties if milestones are not met. Hannerz et al. (2005) identified the need to work long hours in construction to be a significant health risk. Construction is also subject to considerable uncertainty and unforeseen events, often beyond workers' control, that can significantly disrupt production. The competitive, project-based nature of the construction industry creates concerns about job security, with many workers employed on short-term contracts on a project-by-project basis. Combined, these characteristics make construction a stressful industry for workers, whether they perform managerial/professional or manual/non-managerial roles. Furthermore, high levels of work-related effort:

   (i)   reduce opportunities for leisure and recovery;
  (ii)   are associated with disrupted sleep patterns and fatigue; and
 (iii)   have the potential to negatively impact health and wellbeing (van Hooff et al. 2007).

Effort expenditure without sufficient recovery has adverse health consequences, and construction workers report mental health complaints associated with insufficient opportunity to recover from the physical and psychological demands of work (Boschman et al. 2013). Geurts and Sonnentag (2006) also describe how sustained exposure to work demands resulting from working very long hours reduces recovery opportunities, ultimately resulting in chronic health impairment. Reduced opportunities to engage in leisure activity, and preoccupation with work concerns during weekend breaks, are also linked to diminished general wellbeing and performance the following week (Fritz and Sonnentag 2005).

Construction workers' health complaints increase with advancing age. Compared to younger workers, older workers are more adversely affected by psychosocial job demands, including working under time pressure, a lack of employment security, and a concern about unfavourable changes in the work environment (De Zwart et al. 1999). However, although age is a significant factor in workers' health, it may not be the most important. Arndt et al. (2005) note the incidence of work disability in construction workers increases with age, but the dose–response relationship between work exposure to health risks and work disability persists even when age is controlled. They conclude that work-related causes of work disability outweigh age in importance.

4.5 The Management of Occupational Health

Ringen and Englund (2006) describe how it is very difficult to determine construction workers' levels of exposure to health hazards. One challenge associated with making precise estimates of the occupational health risks lies in the fact that exposures are difficult to measure and vary significantly between jobs, within jobs, and over time (Järvholm 2006). Even when attempts are made to measure the exposure of workers to hazardous substances during common construction work tasks, there is significant variation in exposure measurements. Indeed, the range of measured exposures varies as much as 50-fold. This variation and uncertainty increases the need for surveillance and monitoring research to better understand the extent and effects of exposure to health hazards in construction. Given that it is extremely hard to know the extent to which construction workers are actually exposed to health hazards, it would be prudent to exercise the precautionary principle; that is, reflecting the view that it is better to be safe than to be sorry. However, as Ringen and Englund (2006) also point out, the measured exposure levels for common construction work tasks can be well above recommended levels of exposure. If this is the case, then construction workers are likely to be routinely working in ways that could make them ill.

Despite the prevalence of occupational health risks in construction, the industry's health and safety management efforts remain heavily focused on preventing acute effect accidents; that is, the focus is on safety rather than health issues. There is a need to systematically identify and manage occupational health hazards. If these hazards cannot be entirely eliminated, then efforts should be made to reduce the risk to workers' health as much as possible.

Unfortunately, many occupational health risks in the construction industry are managed using lower level behavioural controls. Neitzel and Seixas (2005) note the reliance on hearing protection devices as the primary preventive measure for noise-induced hearing loss. The effectiveness of hearing protection devices is highly dependent on the consistency with which they are used, and construction workers' use of hearing protection devices may be very low (Neitzel and Seixas 2005). Wherever possible, alternative, upper level controls that make the work environment safer (rather than relying on workers' behaviour) should be sought for occupational health risks. For example, Suter (2002) suggests much can be done to reduce noise emissions from construction plant and equipment. Noise exposure during many construction activities can be reduced significantly by considering noise emissions when plant is selected for purchase or hire, having a robust maintenance programme, retrofitting older models with noise reduction devices, and enclosing or insulating the cabins of mobile plant.

Construction workers are often exposed to health hazards arising from the products and materials they use. Where possible, processes involving hazardous substances should be eliminated, and hazardous substances substituted for less hazardous ones. However, manufacturers and suppliers of construction products also have a role to play in driving occupational health improvements. Many construction products are manufactured by suppliers who operate in multinational markets. An international approach to addressing occupational health issues is needed because manufacturers and suppliers of construction products will be reluctant to adopt more stringent precautions in one country than is the required by the norm of all countries within which their products are sold (Ringen and Englund 2006). Case Example 4.2 describes one example of an international effort to reduce risks associated with using epoxy products in construction.

Case Example 4.2 International Initiative to Reduce Risk Associated with Epoxy Products

The Dutch health agency Arbouw teamed up with the European Agency for Safety and Health at Work, the UK Health and Safety Executive, the Bau-Berufsgenossenschaften (Germany), and the Aalborg BST Centre (Denmark). The partners developed an international Code of Practice for working with epoxies, as well as exploring the feasibility of a harmonized ranking system for the health risks posed by epoxy products.

The code emphasized implementing upper level controls, such as substituting epoxy products with less hazardous materials: for example, using cement-based tiling adhesives or silica-based fillers instead. Providing appropriate tools to reduce the risk of epoxy coming into contact with workers' skin was also identified as an important control method. Thus, attaching splash protection shields to rollers and providing spatulas with long handles were recommended. The supply of epoxy kits with well-defined mixing ratios to avoid the need for measuring, and pierce-able dual packs that enable mixing within the pack itself, were also identified as measures that could also reduce the likelihood of skin contact.

The code specified good practices, including allowing epoxy on tools to cure and then scraping it off rather than removing it with solvent, using disposable tools, and closing used epoxy packages immediately. The use of protective gloves at all times was identified as an essential measure. It was recommended that heavy duty gloves made of nitrile, neoprene, or butyl rubber be worn over thin cotton ones, and be used only once.

The classification system for epoxy products ranked them according to the health risks they pose. This classification system was developed in consultation with suppliers of epoxy products. It was expected such a system would help users to select the safest products available, and encourage manufacturers to develop new products posing lower levels of risk.

(Developed from Spee et al. 2006)

4.6 The Health of Construction Workers

4.6.1 The Health Profile of Construction Workers

In a recent study of construction workers in the Australian state of Queensland, data were collected from 90 manual/non-managerial construction workers using the SF-36, a generic, multipurpose short-form survey that produces a profile of health and wellbeing (Ware 1999). Results are shown in Figure 4.2.

The data showed that construction workers had lower levels of mental health than the general Australian population, but slightly higher scores for physical health.

The SF-36 provided a finer-grained measure of health for the construction workers in the sample. Scores could be broken down into various health domains, including physical functioning, role-physical, bodily pain, general health, vitality, social functioning, role-emotional, and mental health. Table 4.1 shows scores for each health domain by age. Scores shown in bold are lower than the population scores for males in the general Queensland population in comparable age brackets. High scores indicate better health. The results show that, with regard to bodily pain, construction workers in

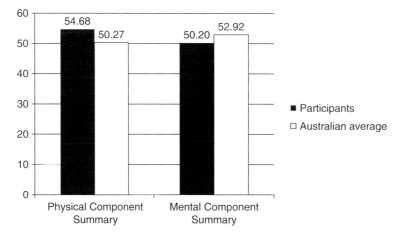

Figure 4.2 Manual/non-managerial construction workers' health scores relative to the Australian population. *Source:* Lingard and Turner 2015.

Table 4.1 Health domain scores by age.

Health domain	Under 30	30–39	40–49	50–59	60 and over
Physical functioning	96.4	93.3	95.5	83.8	80.0
Role-physical	**88.2**	**84.1**	93.5	86.2	90.6
Bodily pain	**79.9**	**72.6**	78.7	**70.8**	79.5
General health	**72.8**	**61.3**	71.3	**63.1**	77.2
Vitality	**62.6**	**54.1**	**63.8**	**51.8**	59.3
Social functioning	**82.5**	**71.7**	**83.4**	**80.0**	100.0
Role-emotional	90.0	**80.2**	91.0	87.3	95.8
Mental health	78.4	**68.2**	76.5	**72.3**	82.5

Note: High scores reflect better health. Bolded figures indicate that the health domain score for construction workers is lower than the equivalent Australian male age-based score.

the younger age brackets are generally less healthy than the average male of a comparable age.

In all age brackets, construction workers reported higher levels of physical functioning than the equivalent male population scores. The findings indicate that younger construction workers (that is, under 30s and people aged 30–39) report lower levels of health than the general population in several domains, including role-physical, bodily pain, general health, vitality, and social functioning. Vitality and social functioning scores among construction workers were low relative to population data workers in all age brackets, except the oldest (60 years and over). Construction workers aged 30–39 reported lower levels of health than the population for all health domains, except for physical functioning.

These findings indicate some differences between the experiences of construction workers and males in the general population, as well as variation in experience by age and between health domains. It is noteworthy that two of the health domains in which construction workers report relatively low scores (vitality and social functioning) reflect health aspects related to long hours and work interference with non-work life. Vitality relates to energy levels and fatigue, and social functioning relates to the extent that physical or emotional problems impact on social activities. It is unclear what reasons produce the apparent difference between the experience of social functioning and vitality of construction workers and that of the general male population. However, project-based construction work involves long and non-standard hours. Previous reviews of the international literature have shown long work hours are related to subjectively reported physical ill-health and fatigue (van der Hulst 2003). Long hours and non-standard hours have also been linked to work–family conflict and burnout in the Australian construction industry, particularly among workers with dependent care responsibilities (Lingard and Francis 2005a).

To understand the progression of construction workers' health over time, data from the *Household, Income, and Labour Dynamics in Australia* (HILDA) survey were analysed. The HILDA survey is a household-based panel study which began in 2001. It collects information about economic and subjective wellbeing, labour market dynamics, and family dynamics. Interviews are conducted annually with adult members of each household and panel members are followed over time. There is a limited number of construction workers in the HILDA dataset and many of these workers have not completed multiple waves of the survey. However, we were able to identify more than 200 participants who had completed five consecutive annual waves of HILDA survey data collection whose industry classification was building construction, heavy construction, and civil engineering or construction services,[1] and whose occupation classification was technical trade worker or labourer.[2] We examined the health domain scores of these workers over five consecutive years and the results are presented in Figure 4.3.

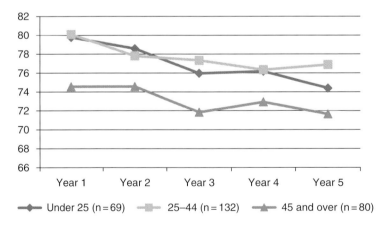

Figure 4.3 General health domain scores by age and year.

1 According to the 1292.0 – Australian and New Zealand Standard Industrial Classification (ANZSIC), 2006 (Revision 1.0).
2 According to the 1220.0 – Australian and New Zealand Standard Classification of Occupations (ANZSCO), 2013 (Version 1.2).

Average general health domain scores were lower for construction workers aged 45 or over at the time data collection commenced. Average general health scores for workers also deteriorated over the five progressive data collection waves, irrespective of their age when data collection commenced.

4.6.2 Mental Health

Construction workers are a high-risk group for mental illness (Doran and Ling 2014). The incidence of mental distress among construction workers is reported to be twice the level of the general male population (Borsting Jacobsen et al. 2013). Peterson and Zwerling (1998) similarly report construction workers experience a significantly higher incidence of emotional and psychiatric disorders than other manual, non-managerial workers in other industries. A review of the HILDA dataset showed the average mental health domain scores for Australian construction workers were more varied and showed different patterns for different age groups. These are shown in Figure 4.4. Among the oldest construction workers (those aged 45 or over when data collection commenced), average mental health domain scores deteriorated over the three waves of data collection and then began to increase in Years 4 and 5. Workers aged 25–44 when data collection commenced reported the highest scores for mental health in the third wave of data collection, but their average scores fell in Year 4 and remained at a similar level in Year 5. The average mental health domain scores for workers in the youngest age group (under 25) at the commencement of data collection gradually increased between Year 1 and Year 4, but fell quite dramatically in Year 5. After five years of data collection, construction workers in the youngest age group when data collection commenced had the lowest average scores for mental health of all age groups.

Much research on the mental health of construction workers has focused on the experiences of managerial or professional workers (see, for example: Leung et al. 2008; Love et al. 2010). Previous research into the Australian construction industry indicated professional and managerial workers had high levels of burnout (Lingard and Francis 2005a, 2006). Burnout, a syndrome comprising emotional exhaustion, cynicism, and a

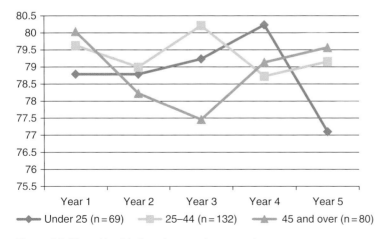

Figure 4.4 Mental health domain scores by age and year.

diminished sense of personal efficacy, has been linked to a range of mental and physical health disorders, and also to unhealthy behaviours (see, for example, Shirom et al. 2005). Recent studies have also found burnout to be very high among manual, non-managerial construction workers. A Dutch study of manual, non-managerial construction workers reports high levels of burnout, leading to early retirement (Oude Hengel et al. 2012).

There is a strong business case for an increased focus on preventive mental wellness promotion strategies in the construction industry, with a A\$2.50 return on every A\$1 invested in mental health programmes, and an even higher return in smaller enterprises (up to A\$15) (PricewaterhouseCoopers 2014). It is argued that work environments promoting mental health will enhance worker wellbeing and happiness, increase engagement and retention, and support greater organizational productivity (Buik and Richards 2015).

4.6.3 Resilience

There is emerging empirical data positioning resilience as an important skill for construction workers. Construction industry workers are known to experience high levels of stress (Campbell 2006; Leung et al. 2008), burnout (Lingard and Francis 2009; Yip and Rowlinson 2006), and work–life conflict (Lingard et al. 2010a). Detrimental outcomes for workers include mental illness, substance abuse, chronic health problems, relationship breakdowns, and intention to turnover. Resilience has been linked to maintaining physical and psychological health, and having the ability to recover more quickly from stressful events (Ryff and Singer 2003). Grant and Kinman (2013) contend that developing resilience enhances wellbeing, job satisfaction, and retention. There is growing evidence that resilience is not an innate, fixed characteristic, but can be developed through carefully targeted interventions (McAllister and McKinnon 2009; McDonald et al. 2013). Developing resilience at work is therefore an area which could be appropriately included in a health promotion programme.

4.6.4 Suicide

Construction is a high-risk industry for suicide. Roberts et al. (2013) report a high incidence of suicide among construction workers in England and Wales when compared to other occupations. The rate of suicide among labourers in building trades was found to be 59.1 per 100 000 (Roberts et al. 2013). This is markedly higher than for the general male population in the UK, with suicide deaths reported to be 9.8 deaths per 100 000 during 2012 (World Health Organization 2016). Similarly, in Australia and the USA, suicide in the construction industry is higher compared with the male general population. In the USA, the construction and extraction occupational category had a suicide rate of 52.5 per 100 000 for males (McIntosh et al. 2016). In contrast, in 2012 the suicide rate per 100 000 of the general male population was 19.4 (World Health Organization 2016). While the USA's construction industry had the second-highest suicide rate when compared with other industries, it had the highest actual number of suicides of all industries (McIntosh et al. 2016). In Australia, suicide rates according to industry are not routinely reported; consequently, the capacity to consider rates in the context of the general population is limited. Using 1995–2001 data from one Australian state, Heller et al. (2007) found the commercial building construction industry had an estimated suicide rate of 40.4 per 100 000 for males, which is higher than the general Australian male population of 16.1 per 100 000 (World Health Organization 2016).

Suicide is a complex phenomenon. It is influenced by a range of interacting factors, including environmental, personal, social, psychological, cultural, and biological factors. Importantly, no single factor is sufficient to explain why a person dies by suicide (World Health Organization 2014). Most commonly, several risk factors act cumulatively to increase an individual's vulnerability to suicidal behaviour (Aleman and Denys 2014; Oquendo et al. 2014; World Health Organization 2014). The workplace has been identified as an environment which can contribute to suicidal intentions for some workers (Oquendo et al. 2014). Harmful psychosocial factors originating from the workplace, and which are linked to suicide, include financial problems, interpersonal conflicts (including bullying and harassment) (Fridner et al. 2009; Nielsen et al. 2015), low control or low decision latitude, low social support, high psychological demands, and long working hours (Hawton et al. 2004; Amagasa et al. 2005; Routley and Ozanne-Smith 2012). Suicide is also more prevalent among males than females (World Health Organization 2014), which means that occupations and industries with male-dominated workforces are more susceptible to higher rates of death by suicide. Furthermore, construction industry workers are known to experience a high level of psychosocial hazards and risks, such as low levels of support at work, low decision latitude and job control, and insecure employment (Alavinia et al. 2007; Oude Hengel et al. 2011; Turner and Lingard 2016a).

Suicide can be prevented (Schwartz-Lifshitz et al. 2012; World Health Organization 2014; Wahlbeck 2015). In recognition of the high prevalence of suicide, there have been calls to increase awareness of suicide and to make suicide prevention a higher priority on the global public health agenda. In response to high suicide rates, the WHO (World Health Organization 2014) released its first report on suicide prevention, stating the report 'represents a significant resource for developing a comprehensive multisectoral strategy that can prevent suicide effectively' (p. 2). In its suicide prevention framework, the WHO (World Health Organization 2014) outlines various strategies which comprise a comprehensive approach. These include:

- raising awareness about mental health, substance use disorders, and suicide;
- gatekeeper training for supervisors and managers;
- education about suicide and its prevention;
- establishing public information campaigns to support the understanding that suicides are preventable; and
- increasing public and professional access to information about all aspects of preventing suicidal behaviour.

The construction industry has responded to the alarming rates of suicide within the workforce by implementing programmes focusing on awareness raising, education, and promoting access to support services and treatments. For example, in Australia, Mates in Construction (MIC) was established in 2008 to reduce the high level of suicide among Australian construction workers. MIC provides suicide prevention through community development programmes on construction sites, and supports workers to access help through case management (Gullestrup et al. 2011). In the UK, Mates in Mind was launched in 2017. It is a sector-wide programme intended to help improve and promote positive mental health and decrease suicide across the construction industry. The programme is led by the Health in Construction Leadership Group (HCLG) and supported by the British Safety Council. In the USA, the Carson J Spencer Foundation released *A Construction Industry Blueprint: Suicide Prevention in the Workplace* (Carson J Spencer Foundation and National Action Alliance for Suicide Prevention 2015). The *Blueprint* has a companion website specially focused on the prevention if suicide in the construction workplace. Some research has begun to evaluate workplace suicide interventions (for example, Milner et al. 2015) and it is important that programmes are rigorously evaluated to assess whether they are effective in preventing suicide in the construction industry.

4.7   The Need to Understand Health Behaviour in Context

4.7.1   Ecological Approach to Health

Health researchers have called for understanding health using an ecological perspective (see, for example, Sallis et al. 2008). An ecological approach tries to understand health in relation to people's contexts, seeking to understand environmental constraints and influences on health behaviour (McLaren and Hawe 2005).

Lingard and Turner (2017) used this approach to understand the environmental constraints that impacted workers' behavioural responses to a health promotion programme introduced at a hospital construction project. They drew on interview data collected from workers and managers to understand the factors shaping health behaviour in the industry, workplace, and family environments. Twenty-two workers were invited to attend a follow-up workshop/focus group to explore their experiences of the health promotion strategies. Another 12 workers participated in one-on-one interviews to gain deeper insight. Participants were asked to reflect on the strategies offered at the project, indicate which strategies they had engaged with, and consider the barriers and supports for engaging in healthy behaviour.

Interview and focus group results are summarized in Table 4.2. The results show some family factors had a positive impact on the adoption of health behaviours by the workers we interviewed. These factors were a motivation to be healthy in order to participate in family life, and also the practical support received from family members to engage in healthy lifestyle behaviours. However, work–family conflict was identified as a barrier to behaviour change. Furthermore, environmental factors identified at workplace and industry levels were overwhelmingly negative in their influence on healthy behaviour. Long hours, time poverty, long commute distances, client demands, fatigue, and job insecurity were all identified as factors that impede adopting a healthy lifestyle. The workers described how the masculine work culture that prevails in the construction industry influences workers' health behaviour, particularly alcohol consumption. The workers also appeared to be resigned to experiencing poor health as an inevitable aspect of working in construction. A similar finding was observed by Kolmet et al. (2006), who describe low health expectations of Australian male manual, non-managerial workers who anticipate they will experience 'wear and tear' caused by the physical demands of their work and inevitable conflict between their work and family lives.

Lingard and Turner (2017) also identified a number of reciprocal relationships that serve to reinforce negative interactions between work patterns, behaviour, and health. Thus, the adoption of healthy behaviours was influenced by workers' experiences of

Table 4.2 Multi-level factors impacting on workers' adoption of health behaviours.

Level of influence	Factors	Example quotations
Family	Work–family conflict (negative)	'My time at home with my wife is limited so I choose to spend it with her. She can't walk as far as I can so we just stay home.'
		Another reflected: 'When you have children it is either you or them. For me to go home and say I want to train [at the gym], that's a big impact on the family.'
	Family support (positive)	'I eat pretty good. My wife cooks my meat and veggies [vegetables] every day and a good lunch. I have fruit in my lunch box every day … might not eat it, but it is there. She looks after me.'
	Family motivation (positive)	'My daughter – as soon as she was born. I do everything for her now. Motivation is for her now. I am a lot calmer since she came along. I used to have a short fuse. Now I have a much longer fuse. She gets up around 5 am and we hang out till I leave for work at 6 am.'
Workplace	Long hours (negative)	'It is a time thing. Some guys go to the gym at 3.30 am in the morning. That is what you have to do in this industry, something has to be sacrificed. Sleep time gets traded. You end up brain dead. We used to have sacred Saturday – do an eight hour day and have Mondays off and then come back to work on a Tuesday. That allowed you to get your jobs done – like going to the doctor or having lunch with your wife, but that got traded and you don't do those things now. If you don't get everything done, it is a sense of underachieving.'
		'Time is the biggest barrier. If you don't have the time, you don't have the time. If you want to do something extra in your day, you will be doing it before you go to work in the dark. By the time you get home, you are exhausted and just want to sit down, you don't want to do anything.'
	Client demands (negative)	'If our client wanted us to build a road in a particular way and they wanted us to invest in the health and wellbeing of our employees and they were able to compensate us through that process to enable us to do that, we would do it and we would make sure we do it because we have to do it. If large-scale clients of that nature make the decision that they want health and wellbeing of the workforce to be a priority then the industry will follow suit … and that comes into work hours, it comes into travel distances, it comes into all sorts of other things.'
	Job insecurity (negative)	'You don't have any job security. For me, when the cranes come down, I don't have a job and work casually till the cranes go back up. Everyone worries. Insecurity affects everyone onsite. The first crane goes by Christmas, those guys start to worry how long before they get work again.'
	Work stressors	'I only had two drinks last week and none so far this week (but it's only Wednesday). But the week before I was having three drinks a night every night (due to the stress).'
		Another worker explained his low level of physical exercise, saying: 'You need to have your head in the right space to exercise.'

(Continued)

Table 4.2 (Continued)

Level of influence	Factors	Example quotations
Industry	Acceptance of injury/poor health (negative)	'I can't do the yoga stretching – I have too many injuries.' 'I don't have time to be healthy – I made it to fifty!'
	Fatigue from physical work (negative)	'If you work in the sun all day with concreting and scaffolding … the last thing you want to do is go home and go to the gym because you have been out there all day. Half the time you don't get lunch. You don't want to be more physical.'
	Masculine work culture (negative)	'… the young guys follow the older guys for years and then make changes. It's the culture on the job site. A lot of guys go straight to the pub.' 'There are over four hundred blokes here. Call it pride but there is no way you going to get me involved with yoga.'
	Travel time (negative)	'When I do long hours of 9.5–10 hours then travel, (while) I'm tired from the physical work I'm also mentally tired. My body wants to get into wind down mode. I'm not ready for another peak from physical exercise.'

Source: adapted from Lingard and Turner 2017.

fatigue and physical tiredness created through working long hours. In particular, this impacted their engagement in physical activity outside work and also led to poor eating habits. However, workers who reported unhealthy eating and exercise patterns also reported deterioration in their levels of physical fitness. In turn, this deterioration increased the impact of work, and increased subsequent levels of tiredness and fatigue. This cyclical effect was reflected in the following comment, which shows how unhealthy patterns of behaviour can become self-perpetuating:

> You get into a cycle. There's not enough time. It's hard to step back and make a change in your lifestyle. You get into a pattern of eat, smoke, drink, sleep. Then you wake up and do it all again. Before you know it you have put on twenty kilos.

The explanations provided by the construction workers in the hospital construction site study suggest environmental conditions at various levels can substantially influence construction workers' health behaviours and experiences (see also Grzywacz and Fuqua 2000). These findings, particularly in relation to workplace factors, are also consistent with research undertaken in the Netherlands that reveals work-related factors (including low levels of job control, high work demands, job strain, a lack of support at work, and ergonomic hazards) are more significantly related to construction workers' health than individual behavioural factors (Alavinia et al. 2007). Similarly, in Sweden, Stattin and Järvholm (2005) found that features of the physical and psychosocial work environment (including physical and environmental hazards, work–life strain, lack of job control, work stress, and high work demands) were stronger predictors of construction workers' experience of musculoskeletal, cardiovascular, psychiatric, and respiratory diseases than individual behaviours.

There is a very real danger that behaviour-based health promotion programmes – as they are currently designed – will draw attention away from important environmental causes of poor health (Chu et al. 1997). Further, if they are introduced without regard to environmental constraints, the impact of these programmes is likely to be weak or short-lived. Understanding the interplay between individual behaviour and work-related risk factors is critical to the design of health promotion programmes (Schulte et al. 2012). Noblet and LaMontagne (2006) call for more comprehensive approaches to the design of workplace health promotion programmes. These programmes should seek to change adverse conditions of work, rather than focus exclusively on trying to change workers' health and lifestyle behaviours.

4.7.2 Interaction Between Work and Family

Australian construction industry workers experience high levels of work–life conflict (Lingard and Francis 2004, 2007; Lingard et al. 2010b). Work–life conflict occurs when 'role pressures from the work and non-work domains are mutually incompatible in some respect' (Greenhaus and Beutell 1985, p. 77). Work–life conflict is linked with negative work-related outcomes, negative non-work-related outcomes, and negative stress-related outcomes. Work-related outcomes include decreased job satisfaction, decreased job performance and intention to turnover. Non-work-related outcomes include a decrease in life and family satisfaction. Stress-related outcomes include depression, burnout, and substance abuse.

Research focused on conflict in the Australian construction industry has investigated the antecedents of conflict. For example, Lingard and Francis (2004) found that workers' experience high levels of work–family conflict was predicted by excessive job demands, including long and irregular work hours. Other investigations into the construction industry have indicated that competitive tendering (MacKenzie 2008) and tight project programming (Lingard et al. 2010c) lead to long working hours, which impact on work–life stress. A further study indicated that hours worked, supervisor support, and work flexibility impacted workers' level of conflict (Lingard et al. 2010b). Research in the Australian construction industry has also indicated that work–life conflict acts as the linking mechanism between work schedule demands and employee burnout (Lingard and Francis 2005b). Additionally, certain job characteristics, such as supervisor support, moderate the relationship between work–life conflict and employee burnout (Lingard and Francis 2006).

While there has been a considerable focus on conflict within the Australian construction industry, some research has investigated work–life interaction from an alternative lens. Some studies have reviewed the barriers to work–life balance and the supports (also referred to as resources) required to enable work–life balance. Turner et al. (2009) found that project culture, resource allocation, and phase of the project were barriers to work–life balance, while project delivery model, flexibility of working hours, and management support acted as facilitators to work–life balance. Work–life balance is described as 'the extent to which an individual is equally engaged in – and equally satisfied with – his or her work role and family role' (Greenhaus et al. 2003, p. 513). Lingard and Francis (2005b) found workers' needs vary according to gender, age, and stage of family, and that work–life supports for workers should move beyond a one-size-fits-all approach and cater for a diverse workforce.

4.7.3 Masculine Work Cultures

Research has also identified social and cultural determinants as relevant to the health of male manual, non-managerial workers. For example, Kolmet et al. (2006) interviewed Anglo-Australian male blue collar workers and found that, although workers are concerned about their health, they also experience a tension between cultural constructs of masculinity (for example, the need to feel 'in control') and their work situations. In construction, employment is rarely secure, work is performed under extreme time pressure, workers often spend significant amounts of time away from their families, and they have little ability to control the way they perform their work. Kolmet et al. (2006) describe how socioeconomic vulnerability experienced by the workers they interviewed created a sense of disempowerment and resignation to the likelihood of diminished life expectancy. Kolmet et al. (2006) describe how cultural constructs of masculinity in work environments, such as construction, negatively impact workers' health. Du Plessis et al. (2013) also describe how 'hyper-masculine' subcultures develop in certain work work environments. In these subcultures, unhealthy lifestyle behaviours are often inadvertently promoted and workers who seek help with health problems are regarded as 'weak' (Iacuone 2005).

4.7.4 Work Ability and Work–Life Fit

The ecological framework contends that all parts of a system are inherently connected, and that the experience of a worker is influenced by the environments in which they are situated. Environments extend beyond the work system and include other systems, such as family, community, and society. An individual's ability to work productively and healthily is inherently shaped and influenced by the demands and resources originating from multiple systems. In instances where demands exceed resources, individuals may suffer from poor health and wellbeing, which may lead to work disability.

Work–life fit refers to a situation in which an individual perceives they have sufficient resources to meet demands arising in multiple life roles, such that their role performance is effective (Voydanoff 2007). Poor fit between work and non-work life is consistently linked to health risk factors including:

(i)   poor diet (Devine et al. 2006);
(ii) high cholesterol (Van Steenbergen and Ellemers 2009);
(iii) lack of physical exercise and low physical stamina (Burton and Turrell 2000; Van Steenbergen and Ellemers 2009);
(iv) high body mass index (Van Steenbergen and Ellemers 2009); and
(v)   harmful levels of alcohol consumption (Frone et al. 1997; Roos et al. 2006).

4.7.5 Health and Work Ability

Construction workers' poor health and experience of work disability are often attributed to risk factors related to lifestyle behaviours and individual biomedical characteristics. For example, Claessen et al. (2009) describe a longitudinal cohort study of construction workers which revealed a body mass index indicating obesity was related, in a follow-up period of approximately 10 years, to occupational disability due to osteoarthritis and/or cardiovascular disease. Similarly, Alavinia et al. (2007) report health status determined by physical health examination of 19 507 Dutch construction workers (including high body mass index, the presence of pulmonary problems, and a 10-year risk for cardiovascular disease) was a significant predictor of workers' longer term ability to work.

There is some evidence to suggest workers' health and occupational disability are best understood as arising from the interplay between occupational risk factors and individuals' health-related behaviours (Van den Berg et al. 2010). For example, Arndt et al. (2005) identify MSDs, cardiovascular disease, and mental disorders as causes of occupational disability among construction workers in Germany, and link these to both occupational risks in addition to so-called lifestyle factors. Oude Hengel et al. (2012) report a combination of occupational and individual factors that are able to predict Dutch construction workers' ability and willingness to work until they reach the pension age (that is, 65 years).

4.8 Organizational Responses to Support Health

In some instances, the work-relatedness of health impacts may be difficult to disentangle from individual biomedical or behavioural risk factors because the links may be indirect and interactive. For example, research indicates psychosocial stress at work is linked to impaired sleep (Åkerstedt 2006). Insufficient sleep is associated with high body mass index and obesity (Bjorvatn et al. 2007; Gangwisch et al. 2005), and insufficient sleep is also an identified risk factor for cardiovascular disease and diabetes (Gottlieb et al. 2006; Gangwisch et al. 2006; Speigel et al. 2005). Thus, the health impact of psychosocial stress experienced at work may, in fact, manifest in health-related behaviour linked to illnesses more commonly attributed to 'lifestyle'.

Workplaces have been identified as environments in which changes can result in significant health improvement through health promotion and disease prevention (Anger et al. 2015; World Health Organization 2008a,b). However, organizational health-related activity is often divided into two distinct areas of practice. These are:

1) occupational health, which refers to the identification and control of known or suspected work-related health hazards; and
2) health promotion or wellness programmes, which refers to the promotion of workers' health not primarily concerned with work-related disease or illness (Pritchard and McCarthy 2002).

Within organizations there is often a distinct separation between these two areas and also an unhelpful confusion between them. While identifying and controlling work-related occupational health hazards is a requirement under occupational health and safety legislation, programmes designed to encourage workers to engage in health-promoting behaviours as a way of preventing chronic illness are not legally mandated. Developing and implementing health promotion and wellness programmes in organizations has become very popular, but it is very important that implementing programmes developed to prevent 'lifestyle' diseases is not seen as a substitute for implementing robust processes for eliminating or reducing exposures to occupational health hazards.

Some workplace health promotion programmes seek to change individual workers' health-related behaviours, while ignoring environmental constraints or factors that impact workers' health (LaMontagne 2004). Programmes implemented in the construction industry have been designed to increase the frequency of workers participating in exercise at home to address shoulder pain (Ludewig and Borstad 2003), and encourage workers to lose weight to reduce the risk of cardiovascular disease (Groeneveld et al. 2010). Research shows these programmes can sometimes produce behaviour change. For example, Sorensen et al. (2007) describe how an individual information campaign produced positive outcomes in smoking cessation and increased consumption of fruit and vegetables in construction labourers in the USA. Gram et al. (2012) also report improved aerobic capacity among Danish construction workers who participated in a work-based physical exercise programme. However, other studies have found little or no significant improvements (see, for example, De Boer et al. 2007; Oude Hengel et al. 2010, 2013). Neither of these large-scale, randomized controlled trials of workplace health promotion programmes produced significant improvements in workers' health or work ability.

In Case Example 4.3 we describe an evaluation of a behavioural health promotion programme introduced in the Australian construction industry. The research was undertaken as part of the Queensland Government's 'Healthier. Happier. Workplaces' initiative (previously Workplaces for Wellness). The initiative was designed to support workplaces to implement programmes that improve workers' health and wellbeing. The scheme specifically focused on five 'SNAPO' health risk factors (that is, smoking, poor nutrition, excessive alcohol intake, physical inactivity, and obesity) (see Begg et al. 2008). The programme's focus was on addressing lifestyle health risk factors in manual, non-managerial construction workers.

Case Example 4.3 Health Promotion in the Australian Construction Industry

A participatory action research (PAR) process was implemented at a large public hospital construction project which was being constructed over four years. The PAR process engaged workers at the sites in generating health promotion strategies. A health promotion planning model was implemented at the sites that included:

(1) undertaking an initial workshop/focus group to identify workers' needs and priorities;
(2) formulating recommendations about health promotion priorities for workers at each site;
(3) monitoring health-related behaviour over time during implementation of health promotion measures; and
(4) undertaking a follow-up workshop/focus group and interviews to explore workers' use and experiences of the health promotion strategies.

Twenty-four workers participated in the initial consultation workshop. They indicated that raising awareness and support for healthy eating would be beneficial because they felt they lacked the knowledge required to make healthy food choices. There was a concern that lack of awareness was perpetuating bad habits, and this was evident onsite. Participants suggested healthy food options be provided at the site canteen, alongside the provision of information about nutrition and healthy food choices. Participants also identified smoking cessation and increasing levels of physical exercise as priority areas for improvement. Given their time poverty, participants suggested opportunities to engage in physical exercise and healthy behaviour during work hours would be most helpful to them.

Case Example 4.3 (Continued)

In response to this workshop, management at the project introduced a series of healthy eating information sessions, onsite yoga and stretching sessions, and a smoking cessation programme. Healthy food options were also introduced to the site canteen.

Workers at the site were asked to complete a weekly log to record their health behaviour over the ensuing weeks. Log data was collected for 13 weeks. The number of logs received ranged from 19 to 99 per week, and the average was 40.

Weekly log data (Figure 4.5) revealed that daily serves of fruit and vegetables and frequency of junk food consumption fluctuated over time. There was a public holiday break during weeks six and seven when workers' junk food intake increased. Workers described how they were 'out of routine' and this influenced their food choices. For example, one explained: '[we are] out doing things that you might not usually be doing – eat a lot of food on the road, do a few trips here and there.' Others actively chose to eat unhealthy takeaway food during time away from work: '[I] want to get out of the routine. Go home for a few beers, then send the kids down the road to get fish and chips.'

During week eight, the site canteen opened with healthy food options. A healthy eating and food tasting session was also held. Daily serves of vegetables increased from week 8 through to week 10. Junk food intake declined but this was not sustained. During week 11, there was a decrease in daily consumption of fruit and vegetables, and junk food was also consumed on more days of the week.

On average, only 10.6 log participants each week indicated they smoked. During week four, a smoking cessation programme was introduced to the site. Workers' self-reported intention to give up smoking increased slightly around this time but this increase in intention to quit smoking was not sustained. One participant described how hard it was for him to give up smoking: 'There have been a lot of grumpy guys getting around

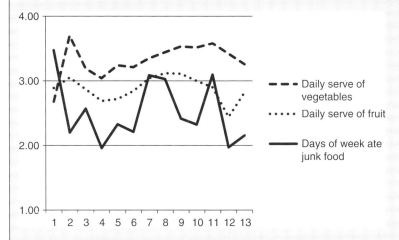

Figure 4.5 Daily intake of fruit and vegetables and frequency of junk food consumption. (1 serve of vegetables = ½ cup cooked vegetables or 1 cup salad vegetables; 1 serve of fruit = medium-sized apple/orange/banana or 2 apricots/kiwi fruit or ½ cup tinned fruit. Junk food is defined as food high in fat, salt, or sugar (such as deep fried foods, hot chips, pies, pastries, chocolates, donuts.)

(Continued)

Case Example 4.3 (Continued)

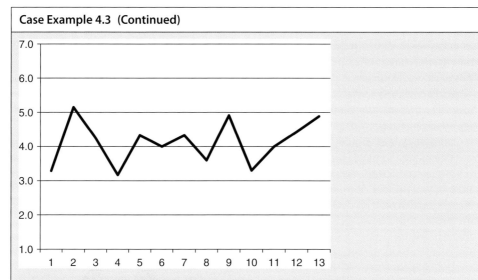

Figure 4.6 Intention of smokers to give up smoking (1 = not at all keen to stop smoking; 7 = very keen to stop smoking).

including myself. I tried (giving up smoking) for a couple of days but it didn't work out, but I had a go and I might have another go' (see Figure 4.6).

Throughout the data collection period, reported levels of physical exercise fluctuated. Physical exercise increased during weeks 2, 6, 7, 9, and 12. During weeks 2 and 12, workers had an extra day off due to a rostered day off. During weeks 6 and 7 there was also a public holiday, providing an extra day off work (see Figure 4.7).

The weekly log data indicated that health behaviours fluctuated over the 13-week period. The data do not indicate that the health promotion measures implemented at the project produced steady or sustained improvements in construction workers' health behaviour.

(Developed from Lingard and Turner 2015)

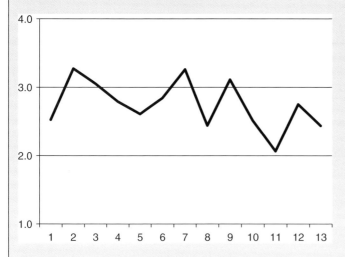

Figure 4.7 Average frequency of physical activity undertaken per week, which is conducted outside work hours and is for 30 minutes or more.

One of the most important lessons to be gleaned from this case example is that there is little benefit in providing one-off, short-term health initiatives. Rather, a sustained and holistic approach is needed. Case Example 4.4 describes the health promotion programme that was subsequently conducted over a 12-month period in the same construction organization. This programme addressed issues identified as being stressful for workers, such as financial literacy, as well as other aspects of health-related behaviour.

Case Example 4.4 A Sustained and Holistic Approach to Addressing Health and Wellness

A health promotion programme was implemented at the Roads Division of a large national construction organization across a 12-month period. The Roads Division consisted of 120 manual, non-managerial workers spread across 18 locations in Australia. The health promotion programme aimed to take a holistic long-term approach to health and safety incident reduction. An initial survey of workers was conducted on seven health and wellbeing areas: mental health, smoking, nutrition, hydration, alcohol, physical activity, and sitting. Results of the survey were used to develop a 12-month health promotion programme that was of interest and importance to the Roads Division workforce. The programme calendar of events was communicated to all workers, and events took place during a 12-month period.

At the end of the 12-month period, a follow up survey was administered to workers to explore the benefits and impacts of the health promotion programme. The most popular workshops were financial literacy, physiotherapy, mental health, breaking bad habits, and healthy cooking demonstrations. Workers spoke of the many benefits they gained from participating in the health promotion programme. The follow up survey identified that 90–100% of respondents:

- *Gained benefit from attending the workshops.* 'Every morning instead of eating a pie, I am having cereal at home.'
- *Shared their learning with family or friends at home.* 'I told the old man to go to the doctors for a prostate/cancer awareness check.'
- *Changed a behaviour or a perspective.* 'I think the workshops are a very positive thing. They might not benefit every person every time, but they do reinforce the fact that the wellbeing of staff is important to the organization. This is helpful for morale. I think we all benefit in some way, no matter what the topic.'
- *Noticed a shift in behaviour and attitudes on their sites.* 'I have noticed that there is a lot more conversations around healthier eating onsite.'
- *Improved health and wellbeing onsite.* 'Every morning at our pre-starts we are taking part in stretches as advised by the physio.'
- *Made the workplace a safer place.* 'I think when workers are treated with respect and consideration, and an overall sense of "family" is created in the workplace, it flows down to making us all care more about the safety of our co-workers. I have been here long enough to see this mob as an extension of my family, and I care about them. I think the workshops help – especially with new employees. It gives them a better sense of the overall attitude of the organisation and its policies towards its workers.'

Workers expressed their interest in participating in future workshops in the areas of mental health, physical activity, bullying, and harassment.

4.9   Conclusions

The construction industry is highly competitive and profit margins are notoriously tight. The capacity to win tenders is the key to survival for construction organizations and significant financial penalties are applied if projects are delivered late. The construction work environment is a challenging environment, consisting of factors including long working hours, interpersonal conflict, isolated work locations, contract-based work, job insecurity, tight work schedules, high workload, and exposure to occupational health and safety hazards. These environmental conditions create ways of working that can be damaging for workers' health, and can create mental and physical disability. An integrated approach to managing worker health suggests a multi-level, systemic approach to managing workers' health is needed. Most importantly, occupational health risks need to be identified and managed, using the most effective methods available.

However, attention also needs to be paid to quality of employment and work in the construction industry. Deeply entrenched ways of working in construction directly impact health, and create environmental conditions that impede workers' ability to engage in healthy behaviours. In this way, the organization of work can contribute to the growing incidence of so-called lifestyle diseases. These environmental constraints also limit the effectiveness of well-meaning (but singularly focused) behavioural health promotion programmes. An integrated model, such as the one presented in this chapter, moves the emphasis away from the behaviour of individual workers and considers health and its various components from a systems perspective. The model highlights that the individual, the organization, the industry, and key stakeholders all in supporting a healthy workforce.

Discussion and Review Questions

1   What type of occupational health issues are experienced by construction workers? Are occupational health issues neglected relative to occupational safety issues in construction? Why or why not?

2   Why might construction industry employers choose to implement health promotion programmes? What advice would you give to a construction organization planning a health promotion programme?

3   What are the challenges to improving construction workers' health? How might these challenges be overcome?

Acknowledgements

This chapter uses unit record data from the Household, Income and Labour Dynamics in Australia (HILDA) Survey. The HILDA Project was initiated by and is funded by the Australian Government Department of Social Services (DSS) and is managed by the Melbourne Institute of Applied Economic and Social Research (Melbourne Institute).

The findings and views reported in this paper, however, are those of the author and should not be attributed to either the DSS or the Melbourne Institute. Some research presented in this chapter was funded by the Department of Justice and Attorney-General under the Queensland Government's 'Healthier. Happier. Workplaces' initiative (previously Workplaces for Wellness) and supported by Lendlease. We also gratefully acknowledge the assistance of Steve Nevin, Risk Manager, Assets and Capital Projects, Fonterra Co-operative Group Ltd.

5

Cultures that Enable Work Health and Safety

Helen Lingard, Rita Peihua Zhang, Nick Blismas, and James Harley

School of Property, Construction and Project Management, RMIT University, Melbourne, Victoria, Australia

5.1 A Culture of Health and Safety

The term 'safety culture' gained prominence because of its use in reports that analysed major safety failures, including the Chernobyl nuclear accident (IAEA 1986), the Piper Alpha oil platform explosion in the North Sea (Hidden 1989), the Clapham Junction rail disaster (Cullen 1990), and other catastrophic events. The inquiries into the causes of these major accidents identified problems inherent in the prevailing organizational cultures which, investigators argued, created the preconditions that allowed these accident scenarios to develop.

James Reason (2000) argues that cultural drivers for health and safety become increasingly significant as health and safety performance improvements 'plateau' following the introduction of safety hardware and software (that is, technologies and systems). The existence (or absence) of a 'safety culture' is frequently referenced by researchers, practitioners, and policy makers when discussing an organization's behaviour or performance. However, as we will discuss in this chapter, the concept of safety culture is poorly defined and somewhat controversial.

In this chapter we identify problems inherent with treating a safety culture as something that:

   (i)   sits aside from the broader organizational culture;
   (ii)   an organization either has or does not have; and
   (iii)   is a panacea for solving health and safety challenges.

We consider two distinct philosophical approaches to understanding culture within organizations and critically consider the management implications that logically flow from these differing perspectives. We identify nine components of an organizational culture that can have an impact on workers' health and safety experiences. These nine components are defined, discussed, and presented in the form of a maturity continuum. The continuum describes characteristics and behaviours of construction organizations operating at five different levels of cultural maturity in relation to these nine components.

Integrating Work Health and Safety into Construction Project Management, First Edition.
Helen Lingard and Ron Wakefield.
© 2019 John Wiley & Sons Ltd. Published 2019 by John Wiley & Sons Ltd.

5.2   What Is Culture and Why Is It Important?

Schein (2010) defines culture as

> … a pattern of shared basic assumptions learned by a group as it solved its problems of external adaption and internal integration, which has worked well enough to be considered valid and, therefore, to be taught to new members as the correct way to perceive, think, and feel in relation to those problems.
>
> *(Schein 2010, p. 18)*

Understanding culture is useful because '[culture] is a powerful, latent, and often unconscious set of forces that determine our individual and collective behaviour, ways of perceiving, thought patterns, and values' (Schein 1999, p. 14).

Culture permeates all aspects of human lives: individuals experience and 'do' culture every moment. Culture guides individuals' behaviours and, in turn, behaviours modify culture (Fellows and Liu 2013). Culture also determines how people communicate and interact with each other, and how people interact with their environment.

In an organization, culture:

- guides decision making and activities at all levels in the organization;
- determines the effectiveness of the whole organizational system; and
- determines efficiency in achieving organizational objectives.

Thus, Schein argues that 'understanding culture can help to explain many of our puzzling and frustrating experiences in social and organizational life' (Schein 2010, p. 7).

Alvesson (2012, p. 166) describes culture as creating 'meta-meanings' that provide clues about how to deal with 'tricky' situations. In an organization, culture can reduce ambiguity by acting as a frame of reference that provides meaningful 'guidelines' about what is important and how to act. Thus, although ambiguity is a common feature of organizational life, bounded ambiguity (expressed through the culture) can create some broadly shared 'rules' about what is acceptable and what is not (Richter and Koch 2004). Consequently, behaviour becomes more predictable and anxiety associated with ambiguity is reduced (Guldenmund 2000).

5.3   Problems Inherent in the Term 'Safety Culture'

Writers on safety culture disagree about whether it should be understood as a 'top-down' or 'bottom-up' phenomenon. Safety culture is sometimes viewed as an 'ideal state' that organizations should strive to achieve. In this view, the safety culture is seen as something that sits aside from the broader organizational culture, something that can be readily manipulated through management intervention and used to support organizational health and safety strategies (Glendon and Stanton 2000).

Implicit in this approach is the assumption that managers should develop a unitary safety culture aligned with managerial ideology and strategy (Glendon and Stanton 2000). Thus, it is assumed that, in an ideal culture, all members of the organization will develop shared ideas and beliefs about health and safety risks and incidents. A top-down perspective on safety culture rarely recognizes that different cultures can coexist within a single organization. If writers taking this perspective do recognize the existence of multiple cultures, they frame such diversity as a cultural weakness because the 'ideal' situation is believed to be a strong and unitary culture in which every member of the organization shares similar beliefs and ideas about what is safe and what is not. Thus, one culture (usually that of management) is seen to be dominant and other cultures, where they are recognized to exist within an organization, are subordinated (Richter and Koch 2004).

This interpretation of the term safety culture has been widely adopted and is implicit in many cultural change programmes focused on safety. However, it is very much grounded in a functionalist view of culture.

The functionalist view of culture assumes the social world is composed of concrete empirical artefacts and relationships which can be identified, studied, and measured using a scientific approach. The functionalist view assumes social change can be achieved through 'social engineering', meaning culture is subject to manipulation by groups in positions of power or authority. In the functionalist view, considerable importance is placed on understanding order, equilibrium, and stability in society, and the way these attributes can be managed. The functionalist view is concerned with the effective 'regulation' and control of social affairs. Those who adopt a functionalist view see culture as being

> … made up of those mechanisms by which an individual acquires mental characteristics (values, beliefs) and habits that fit him [sic] for participation in social life; it is a component of a social system which also includes social structures, to maintain an orderly social life, and adaptation mechanisms, to maintain society's equilibrium with its physical environment.
>
> *(Allaire and Firsirotu 1984, p. 217)*

A functionalist approach views organizational culture as the shared values and norms within the organization, and emphasizes leaders' roles in cultivating the culture through developing managerial ideology, goals, and strategy (Schein 2010). Organizational culture should be strategically managed to serve the purpose of the organization (Waring 1992). It is assumed that organizational cultures can be 'engineered' by identifying their essential components and formulating strategies to develop these components across the organization.

In keeping with the functionalist perspective, many definitions treat safety culture as an entity that an organization either has or does not have (Hale 2000). It is assumed that if an organization has a safety culture then it will perform well in safety, and if a safety culture does not exist then it will perform poorly. For example, the UK Health and Safety Executive's (HSE) Advisory Committee on the Safety of Nuclear Installations offers a widely accepted definition of a safety culture as:

> … the product of individual and group values, attitudes, competencies, and patterns of behaviour that determine the commitment to, and the style and proficiency of, an organisation's health and safety management.
>
> *(ACSNI 1993)*

However, some argue this definition is too narrow because it may not adequately capture all the organizational and social factors that are important to the healthy and safe operation of a workplace (Sorensen 2002).

5.4 Organizational Culture as an Enabler of Safety

There is an alternative and, we suggest, more useful way of understanding the relationship between culture and safety within an organization. This alternative perspective positions safety as an outcome (rather than a subset) of the organizational culture. This view assumes organizational cultures have characteristics that impact on the way health and safety are prioritized and enacted within workplaces. But it moves away from the notion that a safety culture can be 'bolted on' to an organization or easily engineered through managerial intervention.

Guldenmund (2000) argues that the basic assumptions underlying the operation of an organization have a profound impact on the effectiveness with which health and safety are managed in that organization. Safety might be a core value in some organizations, but not in others. It is likely that health and safety activities will be driven by all the basic assumptions that make up the organization's underlying culture, whether these are specially concerned with health and safety or not. Similarly, Antonsen (2009a) writes, 'there is no such thing as a safety culture, but rather there are different traits of larger organizational culture that can affect the organization's safety levels' (p. 184). He argues work-related attitudes and behaviours should be analysed and understood as being situated in a wider organizational context in which organizational culture provides a shared framework of reference for meaning and action. This distinction is also reflected in the difference between whether one views culture as something an organization has, or as something an organization is (Smircich 1983, p. 347).

The latter perspective takes an interpretive approach to understanding culture. An interpretive view seeks to understand the world *as it is*, and to understand the fundamental nature of the social world through subjective experience (Burrell and Morgan 1994). Those adopting an interpretive perspective see culture as developing through an emergent social process, created by individuals. Culture is regarded as a system of meanings and symbols shared between groups of individuals who participate in this social process (Allaire and Firsirotu 1984). The interpretive view suggests culture cannot be shaped or manipulated easily, and cannot be studied easily using scientific methods. Culture does not reside in the attitudes and/or cognition of individuals. It resides in the 'meaning' shared by social actors (Allaire and Firsirotu 1984). Consistent with this view, Geertz defines culture as 'the fabric of meaning in terms of which human beings interpret their experience and guide their actions' (Geertz 1973, p. 145).

In an organizational context, an interpretive approach views culture as an emergent property of the organization, in which. shared meanings and interpretations are created (or re-created) collectively and continually by the members of an organization or organizational sub-unit (Demers 2007). Culture is used to inform beliefs, behaviours and create a sense of collective identity (Naevestad, 2009). People who subscribe to the interpretive approach believe that, because organizational culture is created by all organizational members it cannot be manipulated easily or created by senior managers, and it is not 'owned' by the organization. Thus, the interpretive view represents a bottom-up (rather than a top-down) approach to organizational culture.

The interpretive perspective acknowledges that multiple subcultures may develop within an organization. For example, Gherardi et al. (1998) describe how engineers and construction managers developed differing patterns of meaning about health and safety through a dynamic process of interaction and negotiation.

In the interpretive view, no culture dominates by default. Viewed from this perspective, non-leader-centred sources of culture are recognized as important and influential and differing points of view can be brought together to deal effectively with problems, challenges, and daily organizational frustrations (Blewett et al. 2012). Differentiated cultures have been viewed as the product of various types of social grouping. For example, Parker (2000) describes cultures as forming around three types of social grouping (which can also overlap within an organizational context):

- spatial/functional (for example, buildings, sites, or departments);
- generational; and
- occupational/professional.

Hale (2000) adopts this line of argument, stating it is more appropriate to talk about the (organizational) cultural influences on health and safety, rather than a single, uniform safety culture. Similarly, Haukelid (2008) argues that 'safety culture should not be something separate from – or in addition to – an organizational culture, but constitute an integrated part of this culture' (p. 417). Hopkins (2006a) also distinguishes safety from culture by examining the way in which organizational cultures influence health and safety. Each organization has its own culture, and that culture is expected to influence health and safety.

5.5 Different Approaches to Understanding Cultural Influences on Health and Safety

When defining and understanding cultural influences on health and safety, it is important to consider these opposing views of culture. The view that is chosen will have relevance for:

- what aspects of culture should be considered important to health and safety;
- the choice of strategies that might effectively enhance the cultural influences on health and safety; and
- the way cultural influences on health and safety should be understood or assessed.

For example, the functionalist approach to safety culture has been embedded in the traditions of social and organizational psychology and favours quantitative methods of assessment. Safety culture is seen as an entity that can be measured using tools, such as perception and attitude surveys. On the other hand, the interpretive approach to safety culture is embedded in the traditions of sociology and anthropology and favours qualitative rather than quantitative methods. From an interpretive perspective it is argued that in-depth study, interviews, observations, and document analysis are required to reveal the underlying and shared systems of meaning that members of an organization have about health and safety.

5.6 Multiple Layers of Organizational Culture

According to Schein (2010), confusion arises as a result of the failure to recognize three different layers at which organizational culture operates. He developed a three-layer model of organizational culture (shown in Table 5.1). The differentiation between each layer is based on the 'degree to which the culture phenomenon is visible to the observer' (Schein 2010, p. 23). This model suggests that the basic assumptions at the deepest level of an organization's culture shape the way organization members interpret and interact with the environment around them.

Based on Schein's model, organizational cultures (that can drive health and safety outcomes) have three layers:

- the deepest layer (basic assumptions);
- an intermediate layer (espoused beliefs and values); and
- the surface layer (behaviours and artefacts).

As discussed above, some of the basic assumptions that underpin an organizational culture might not be specifically concerned with health and safety, but they might still have some health and safety impact. For example, Guldenmund (2000) suggests a basic assumption that written rules and procedures are futile is not specifically related to health and safety, but it will influence the compliance of people within the organization to rules and procedures, which could have a health and safety impact.

Table 5.1 The three layers of culture.

Basic assumptions	Usually unconscious, taken-for-granted beliefs and values.
	They are developed over a long period and shape the way group members perceive, feel about, and interpret the environment around them.
	They are the essence of any culture.
Espoused beliefs and values	The principles that guide group members in their behaviours. They include ideals, goals, values, aspirations, and ideologies.
	Espoused beliefs and values consciously held and explicitly articulated because they guide group members in how to deal with certain key situations.
	They are used to train new members in how to behave.
Behaviours and artefacts	Artefacts are symbols that reflect the basic underlying assumptions and espoused beliefs and values of an organization.
	Artefacts include:
	• visible organizational structures (like organizational charters, formal responsibility descriptions, and organizational charts);
	• organizational processes; and
	• observed behaviour that accompanies organizational processes.

Source: Schein 2010, p. 24.

The second layer in Schein's model – that is, espoused beliefs and values – aligns with the 'managerial ideology' emphasized by a functionalist approach to organizational culture. Espoused values and artefacts relate to what managers 'audibly say and visibly do' about organizational goals and aspirations. Schein's choice of terminology reflects the fact that what is seen and heard in an organization is not always a true expression of the underlying culture (Guldenmund 2000).

An example of multiple layers of culture is illustrated in research undertaken by Sherratt et al. (2013) in the UK construction industry. They analysed the way health and safety is written and spoken about at construction sites. Safety signage, safety-related communication with workers, safety manuals, and memos (artefacts in Schein's three-layer model) reflected an 'enforcement' orientation to managing safety. These artefacts reflect a belief that a command and control management style is needed to ensure health and safety compliance (an intermediate level belief in Schein's model). This belief, and the artefacts that flow from it, can be traced to a more basic assumption about the need for external rules and enforcement to regulate behaviour. Sherratt et al. (2013) highlight the ambiguities that arose because the enforcement-oriented organizational culture was sometimes at odds with statements in corporate health and safety policies about worker engagement in, and ownership of, health and safety.

Schein's three-layer model of organizational culture could help to resolve the ongoing debate about how cultures and their impact on health and safety should be analysed and understood.

Guldenmund (2000) argues that basic assumptions reflect the core of an organization's culture, while the two outer 'layers' (beliefs and espoused values, and artefacts and behaviours) are more appropriately described as the health and safety climate. Following Guldenmund (2000), health and safety climate might usefully be viewed as the 'surface' expression of the culture that has the potential to influence health and safety. The distinction between culture and climate, as reflecting layers of varying depth, has been adopted by a number of health and safety culture/climate researchers (for example, Havold 2010).

Schein (2006, p. 14) writes that 'culture is the deepest, often unconscious, part of a group'. Basic assumptions are particularly difficult to identify, as people may not even recognize they have these assumptions, or they appear to be so self-evident that they are not talked about. There is general consensus that health and safety climate surveys cannot reveal the basic assumptions underpinning an organization's culture. Alternative methods are recommended to explore and understand culture at its deepest level (Flin et al. 2000; Guldenmund 2007). Understanding basic assumptions therefore requires qualitative approaches, observing and interpreting organizational members' interactions and behaviours, from which basic assumptions can be inferred. Recognizing this, Fruhen et al. (2013) trialled a method for exposing basic assumptions underpinning a safety culture by analysing managers' language as symbolic behaviour that transmits values, norms, and meaning.

The two outer layers of organizational culture are relatively easy to observe and measure. Artefacts are the tangible products of the organization's espoused beliefs and values. They can be assessed readily using tools such as checklists and activity analyses. It is also acknowledged that workers form perceptions of managerial actions over time and these perceptions are amenable to measurement by employee perception/attitude surveys. Consequently, espoused beliefs and values are frequently measured using questionnaire survey tools (usually referred to as health and safety climate surveys).

However, although health and safety climate surveys provide a 'snapshot' of the organizational environment at a given point in time, the health and safety climate is believed to be relatively unstable and subject to change (for example, as a result of factors in the operational environment). We discuss this in greater detail in Chapter 6 in relation to measuring health and safety and the use of health and safety metrics.

In contrast, the basic assumptions underlying an organization's culture are viewed as relatively enduring characteristics reflected in a consistent manner of dealing with health and safety issues. Wiegmann et al. (2004) suggest the organizational culture compares to the personality of the organization, while the health and safety climate compares to the mood of the organization at a particular point in time. The state of the health and safety climate ascertained using questionnaire survey tools can provide important information about *what* is happening in an organization at a particular point in time, but understanding the culture is required to explain *why* health and safety is enacted in a particular way (Borys 2012a).

Consequently, critics of health and safety climate surveys suggest that they merely 'scratch the surface' of culture and that a broader suite of methods is needed to understand culture fully. It is recommended that climate/culture surveys be understood in the context of the social processes and meaning attributed to practices and events by people in the work environment (Antonsen 2009b).

5.7 Understanding Cultural Influences on Health and Safety

The discussion so far in this chapter of cultural influences on workers' health and safety suggests some important considerations. These are summarized in Sections 5.7.1–5.7.4.

5.7.1 Rethink the Way Culture Influences Health and Safety

It has been unhelpful to treat health and safety culture as something an organization either has or does not have. Health and safety culture does not exist as an entity separate from the broader organizational culture. It cannot be engineered or 'bolted on' to an organization to improve the effectiveness with which health and safety is managed. Blewett (2011) advocates removing use of the words 'health' and 'safety' in association with 'culture'. It is recommended that researchers consider the organizational culture and the way this culture influences health and safety, rather than referring to health and safety culture and assuming this is a distinct and separate subset of the organizational culture. For this reason we prefer to focus on aspects of an organizational culture that enable (or impede) health and safety performance. We return to this point in Section 5.8 in our discussion of the Australian Constructors Association's Organisational Culture Maturity Continuum.

5.7.2 Understand Culture as a Layered Phenomenon

The terms 'culture' and 'climate' are often used interchangeably, particularly when used in association with health and safety (Cox and Flin 1998). Mearns and Flin (1999) call for a clearer distinction between the concepts of organizational culture and safety climate, arguing that using the terms interchangeably causes misunderstanding and confusion. Understanding cultural influences as being multilayered is helpful in distinguishing between the underlying organizational culture and the health and safety climate that prevails at a given point in time. Understanding the difference between culture and climate in terms of their depth and stability is particularly useful in the dynamic, constantly changing, construction project environment in which operational issues might produce short term fluctuations in the safety climate.

5.7.3 Adopt a Multimethod Approach

The basic assumptions underlying organizational cultures (often arising from past events) are best exposed through using qualitative methods of field research or ethnography, which provide rich information about the organization's value system. Climate surveys measuring workers' perceptions of health and safety in a workplace can usefully measure the perceived effectiveness of changes in organizational health and safety practices. They provide reliable and valid information about what is happening in an organization or project but do not answer questions about why health and safety are enacted in a particular way (Borys 2012a). A deeper understanding of the organizational or project culture would require a more qualitative investigation of the way managers' and workers' behaviours are influenced by the unconscious, taken-for-granted beliefs and values that team members bring to a project.

5.7.4 Appreciate Culture as a Differentiated Concept

The interpretive perspective regards culture as the shared meaning that naturally emerges through interaction between members of a social group. This approach acknowledges that multiple cultures can coexist and that non-leaders in organizations can be a source of culture. In the organizationally complex construction industry, a pluralistic approach to understanding cultural influences on health and safety is likely to be helpful. Projects are delivered through temporary multidisciplinary teams. Each organization involved in a project will have its own organizational culture, and team members will bring their assumptions, beliefs, and values to the project. But a distinct project culture may also emerge as a product of the social interactions between team members over the life of the project. The impact on organizational cultures is unclear when teams disperse and members return to their employing organizations. Further research is needed to understand the relationship between organizational and project cultures in the fragmented construction industry.

5.8 The Australian Constructors Association's Cultural Maturity Model

The RMIT Centre for Construction Work Health and Safety Research was tasked with undertaking a review of research relating to organizational culture and workers' health and safety. The outcome of this review was the identification of nine components of organizational culture that influence health and safety. These components are explained below.

5.8.1 Component 1: Leadership

Managerial behaviour is recognized as a key aspect of organizational culture with the potential to impact workers' health and safety. When managers clearly and explicitly express their strong health and safety values, and reinforce these values with consistent behaviour, then adopting safe and healthy work practices is more likely to be regarded as the unconditional 'way of doing things' in the workplace.

O'Dea and Flin (2001) identify participative leadership as particularly important in developing a culture that enables health and safety. There are four facets of participative leadership, identified in Table 5.2.

A transformational leadership style has also been linked to good health and safety outcomes (Barling et al. 2002; Zohar 2002b). Transformational leaders demonstrate the following characteristics:

- idealized influence;
- inspirational motivation;
- intellectual stimulation; and
- individualized consideration (Kelloway et al. 2006).

Zacharatos et al. (2005, p. 80) suggest four ways in which transformational leadership enhances health and safety performance. These are summarized in Table 5.3.

Mullen and Kelloway (2009) provide evidence that developing safety-specific transformational leadership capability in managers improves health and safety in workplaces. In contrast to the positive effect of transformational leadership, Kelloway et al. (2006) report negative impacts on performance when health and safety leadership is passive or

Table 5.2 Four facets of participative leadership.

1. Visibility	Effective leaders:
	• are visible;
	• participate in health and safety activities at the workplace;
	• consistently apply health and safety policies and rules;
	• model good health and safety practices; and
	• lead by example.
2. Relationships	Effective leaders:
	• form open, honest relationships with the workforce by engaging in two-way communication; and
	• listen and respond to workers' suggestions for health and safety improvements.
3. Workforce involvement	Effective leaders:
	• actively involve workers in work planning and decision making.
4. Proactive behaviour	Effective leaders:
	• proactively seek to improve health and safety; and
	• promote an environment in which hazards and incidents can be reported without fear of reprisal.

Table 5.3 The influence of transformational leadership on health and safety.

1. Leaders high in idealized influence convey the value of health and safety through their personal experience.

2. Leaders high in inspirational motivation can convince their followers that high levels of health and safety, not previously considered possible, can be achieved.

3. Intellectually stimulating leaders help followers think about health and safety and develop new ways to achieve high health and safety levels.

4. Individualized consideration is evident through leaders' real concern about their followers' health and safety at work.

laissez-faire – for example, failing to intervene until problems become serious enough to require attention or delaying decision making.

Zohar (2002b) also differentiates between transactional and transformational leadership, suggesting:

- transactional leadership provides reliability and predictability ('expected performance'); while
- transformational leadership provides heightened motivation and development orientation ('performance beyond expectations').

Both transformational and transactional leadership are probably important to ensure optimal health and safety performance. However, leadership that reflects a greater concern for workers' welfare and closer, individualized, relationships creates stronger and more positive group safety climates and reduced incidence of injury (Inness et al. 2010; Zohar 2002b).

Consistency is an important characteristic of managerial leadership behaviour in relation to health and safety. This is highlighted by Mullen et al. (2011), who report managers do not always demonstrate the same style of leadership. However, when managers alternate between transformational and passive leadership behaviours, they minimize any positive effects of transformational leadership behaviour on workers' health and safety. The key messages are that:

- it is insufficient to promote health and safety occasionally; and
- to produce a positive influence on health and safety performance, transformational leadership needs to be consistent.

Recent research highlights the need to evaluate the quality of health and safety leadership at different levels within an organization. Transformational leadership is likely to be important at all managerial levels. However, Flin and Yule (2004) suggest managers at different levels should engage in different types of leadership behaviour, as shown in Table 5.4.

There are practical reasons for evaluating health and safety managerial leadership behaviour at different levels within an organization. Senior managers play a key role in establishing an organization's health and safety policy, setting strategic objectives for health and safety, and allocating organizational resources to the overall management of

Table 5.4 Different levels of managerial influence on health and safety.

Senior managers	Senior managers effectively set the 'tone' of health and safety activity within an organization. Their transactional leadership includes allocating resources to the management of health and safety and ensuring the organization's health and safety management programme and processes are compliant and effective.
	Transformational leadership at a senior management level can involve continuously (and visibly) demonstrating a strong commitment to health and safety. This is best demonstrated by devoting time to health and safety matters within the organization and encouraging lower level managers to adopt a participatory management style with regard to health and safety.
Middle managers	Middle managers can demonstrate transactional leadership by ensuring effective health and safety communication and compliance with organizational health and safety systems.
	Transformational leadership at a middle management level can involve communicating organizational health and safety goals and values to supervisors and workers and emphasizing health and safety in the context of schedule or production pressures.
Supervisors	At a supervisory level, transactional leadership styles are likely to be effective when they focus on monitoring compliance and reinforcing health and safety practices. Transformational leadership at a supervisory level can involve encouraging workers' participation in health and safety activities and actively supporting organizational health and safety initiatives.

Source: adapted from Flin and Yule (2004).

health and safety. However, workers 'at the coalface' have little direct contact with senior management. Consequently, the role played by middle managers and supervisors is critical (Zohar 2002a). This is particularly the case in decentralized, project-based industries like construction. Supervisors are particularly influential because they 'filter' organizational health and safety messages. Put simply, supervisors communicate what 'management really wants'. Our research shows that, in construction projects, first level supervisors have very strong, direct influence on local health and safety behaviour and performance (Lingard et al. 2010b, 2012b).

5.8.2   Component 2: Communication

Open, frequent, and multidirectional communication about health and safety is identified as an important component of an organizational culture that enables health and safety performance (Health and Safety Executive 2005a,b). Health and safety communication serves to:

- inform workers about hazards, risks, and ways of working safely;
- elicit important information about workers' experiences and concerns; and
- elicit suggestions for ways to improve health and safety.

The UK Health and Safety Executive (2005a) suggests effective health and safety communication within an organization occurs in three directions:

- top-down – management to frontline;
- bottom-up – frontline to management; and
- horizontal – between peers or functional groups.

Top-down communication ensures health and safety goals and objectives are under-stood by workers, and that health and safety-related information is transmitted to employees in a timely way. It is mainly concerned with:

- passing on health and safety policies and statements;
- disseminating information related to risks and safety, such as hazard analysis and preventive measures; and
- providing feedback to respond to workers' reporting and raising health and safety concerns.

Bottom-up communication is mainly concerned with reporting and issue-raising, by which workers report health and safety issues and concerns to management for action and improvement. Olive et al. (2006) suggest organizations should develop an atmos-phere (and supporting structures) that allows workers to feel comfortable asking ques-tions or raising concerns, or making suggestions about health and safety procedures or ways of working. This can help to minimize latent shortcomings in a work system by identifying and resolving problems when and as they arise and ensuring that work pro-cedures are developed with practical input from the people who do the work. The ben-efits to be gained from participatory management approaches, in which workers are effectively consulted in the design of work are discussed further in Chapter 7.

Horizontal communication refers to the transfer of health and safety information between peers, departments, and functional units. This is important when technical and organizational elements need to be coordinated to manage health and safety issues. In a construction project, coordination is extremely important between trades and between functional groups, such as people responsible for various aspects of technical design work for a facility. The link between communication within design teams and health and safety outcomes is described and discussed in Chapter 3.

Communication can either be formal or informal. Informal communication enables managers to verbally communicate the importance of health and safety and to listen to workers' concerns. Examples include conducting management tours and 'walking the job, talking to people, listening to people' (Health and Safety Commission 2001, p. 67). Managers can develop a deeper understanding of health and safety issues by actively discussing challenges and issues with workers.

Relationships are critical to effective communication. Good supervisor–employee relationships are conducive to workers' willingness to raise safety concerns with their supervisors. Where relationships are good, workers are:

(i) more likely to raise health and safety concerns and internalize the organiza-tion's health and safety values; and
(ii) less likely to be involved in a work-related accident (Kath et al. 2010; Mullen 2005).

Open and honest communication also engenders trust. Conchie and Burns (2008) investigated the effects of open communication on workers' belief and trust in an organization's risk management processes. They report that open communication about health and safety risks significantly contributes to workers' trust in risk manage-ment processes and decisions.

5.8.3 Component 3: Organizational Goals and Values

What is valued and what the organization and its members aspire to be are fundamentally shaped by the basic assumptions at the heart of organizational culture. Guldenmund (2000) has pointed out that an organization's core cultural beliefs and assumptions do not have to be especially concerned with health and safety. They can be about any number of things that may or may not have an impact on health and safety.

In the organizational environment, protection of workers' health and safety can be seen as a conflicting goal when seen alongside production – at least in the short term (Reason 1998). Balancing these conflicting imperatives can be delicate and difficult. Reason (2000) suggests the way in which conflicts are resolved and trade-offs are made reflects the underlying organizational culture. He uses the introduction of the Davy lamp to the mining industry in the 1800s to illustrate the paradoxical nature of the protection–production trade-off. To reduce the risk of explosions in mines, the Davy lamp was introduced to isolate the light source (a naked flame) from combustible gases. However, mine owners recognized using the Davy lamp enabled miners to work in rich coal seam areas previously considered too dangerous to mine. Ironically, following the introduction of this new protective technology, the incidence of mine explosions increased.

Analysis of serious organizational accidents often reveals the existence of cultural drivers that 'normalized' unsafe practices and led people to ignore early warning signs in order to maintain production or project progress. For example, Hopkins (2006a,b) described a situation in the rail industry in which a culture of punctuality in running trains resulted in denying risk in the operating environment, culminating in a serious accident. Hopkins (2006a,b) also documented how a culture in which production was valued more highly than safety – a 'can do' attitude and a command and discipline orientation – created the conditions in which a number of Air Force personnel were exposed to toxic chemicals over a 20-year period. The cultural assumption that a high production rate is for 'the greatest good' of the organization is often cited as a factor in health and safety corner-cutting (see, for example, Guldenmund 2000). In the construction industry, time, and cost are so ingrained as basic assumptions about what constitutes a successful project that it is easy to imagine negative health and safety impacts arising in organizations with less mature cultures.

In many situations, the basic assumptions driving organizational behaviour are not specifically concerned with health and safety. However, a belief in the importance of health and safety can be one of an organization's basic assumptions. Arguably, this will create the conditions required for health and safety to be prioritized within the organization in the context of competing organizational objectives.

Zwetsloot et al. (2013a) have proposed that health, safety, and wellbeing at work represent important values in themselves. However, other organizational values (or 'basic assumptions') also contribute to health and safety outcomes. They identified three clusters of organizational values that are influential to health and safety in an organization. These are summarized in Table 5.5.

5.8.4 Component 4: Supportive Environment

Various features of the physical and psychosocial work environment influence health and safety-related behaviour and performance (Christian et al. 2009). Having a

Table 5.5 Organizational values and their influence on health and safety.

Value	Influences
Valuing people	A positive attitude towards people and their 'being', including core values of interconnectedness, participation, and trust
Valuing desired individual and collective behaviour	'Doing', primarily comprising core values of justice and responsibility
Valuing alignment of personal and organizational development	'Becoming', characterized by core values of development, growth, and resilience

Source: (Zwetsloot et al. 2013a,b).

supportive work environment is believed to influence health and safety directly, because it results in open and effective communication and appropriate levels of training, resource allocation, work planning, and supervisory concern for health and safety.

However, organizational support is also believed to influence health and safety indirectly by engendering higher levels of organizational commitment (Barling et al. 2003), job satisfaction (Parker et al. 2001), and trust (Zacharatos et al. 2005).

A great deal of research has focused on perceived organizational support – that is, the global perceptions workers form about the extent to which the organization they work for is concerned about their wellbeing (Eisenberger et al. 1990). Perceptions of organizational support have been linked to workers' compliance with organizational health and safety policies and reduced involvement in work accidents (Gyekye and Salminen 2007). Wallace et al. (2006) used the term 'foundation climate' to describe workers' perceptions of the ambient climate for organizational support and management–worker relationships. They found that the perceptions of support (expressed in the foundation climate) were strong predictors of safety outcomes. Similarly, Larsson et al. (2008) reported that when construction workers have favourable perceptions of their work environment (in terms of the psychosocial conditions experienced at work, including social support) they are more likely to demonstrate positive interactive and personal safety-related behaviour.

Work organization has also been examined as a driver of health and safety outcomes. Work organization refers to the 'way work processes are structured and managed, such as job design, scheduling, management, organizational characteristics and policies and procedures' (DeJoy et al. 2010, p. 140). Various aspects of job design have been linked to better safety performance – including job autonomy (Parker et al. 2001; Barling et al. 2003), task variety, and opportunities for skill development (Barling et al. 2003).

Zacharatos et al. (2005) examined the relationship between health and safety and high performance work systems. They identified 10 features of a high performance work system linked to workers' personal safety orientation and fewer safety incidents (see Table 5.6).

The psychosocial work environment has been linked to workers' mental health and wellbeing, as well as safety; for example, Nahrgang et al. (2011) report a supportive work environment is the most consistent predictor of workers' burnout, engagement, and safety outcomes.

Table 5.6 Zacharatos et al.'s 10 features of a high performance work system.

Feature	Descriptor
Employment security	The extent to which an organization provides stable employment
Selective hiring	Ensuring a fit between workers and the work environment
Extensive training	Allowing workers to acquire competencies to control their work
Self-managed teams and decentralized decision making	Fostering cohesion and a sense of safety responsibility
Reduced status distinctions	Increasing communication between managers and workers
Information sharing	Ensuring people have full information required to perform their work
Compensation contingent on safe performance	Paying people well and recognizing safe working practices
Transformational leadership	Providing a stimulating, motivational, and caring work environment
High quality work	Including appropriate workload, role clarity, and job control
Measuring management practices	Ensuring the quality of the organization's health and safety effort is appropriately measured

5.8.5   Component 5: Responsibility, Authority, and Accountability

Clearly articulated and understood responsibilities for health and safety are a feature of organizations with good health and safety performance. In the construction industry Törner and Pousette (2009) report that attainment of high health and safety standards requires people at many levels in an organization to assume responsibility for health and safety in their work. Managers need to allocate resources to a level consistent with and sufficient to meet the organization's health and safety objectives. This includes allowing sufficient time for people to perform their work safely. Adequate 'thinking time' is needed so workers can plan and carry out their work in a safe and healthy manner (Glendon and Litherland 2001). Pre-start sessions with supervisors play a key role in preparing workers for their daily tasks. The proactive resolution of conflicts between safe working practices and schedule-driven pressures is characteristic of enabling health and safety cultures (Health and Safety Executive 2012) and effective planning and pre-start communication between workers and supervisors can facilitate this.

Responsibility for health and safety is not held exclusively by managers. There is a growing recognition that co-workers have a role to play in looking out for, and helping to protect, the health and safety of their workmates. For example, Burt et al. (1998) developed the Considerate and Responsible Employee (CARE) scale to measure workers' attitudes towards their co-workers' safety. The CARE scale comprehensively covers various aspects of a caring attitude, including:

- reminding co-workers about hazards;
- assisting co-workers to work safely;
- discussing and sharing safety information with co-workers;
- correcting co-workers' unsafe acts;

- avoiding creating hazards to co-workers by their own behaviours; and
- informing management about hazards.

Burt et al. (2008) found that workers' willingness to intervene to protect the safety of their co-workers is linked to their trust in managers' commitment to workplace health and safety. When the organizational culture is characterized by trust and open communication, workers will be much more likely to stop their own work to help a co-worker or inform their supervisor of any concerns they have about the safety of themselves or their co-workers.

However, responsibility must be accompanied by accountability and authority. It is important that people are not punished for actions, omissions, or decisions taken by them which are commensurate with their experience and training and which make sense in the organizational context in which people are working. (The issue of why people break rules is dealt with in greater detail in Chapter 7.) At the same time, however, people do need to be held responsible for acts of wilful misconduct or negligence. The concept of a 'just culture' captures the need for balanced accountability, applying to individual workers, managers and other parties responsible for designing work processes and systems of work (Dekker 2008).

5.8.6   Component 6: Learning

Learning is a vital component of an organizational culture that enables workplace health and safety. Reason (1997) describes a learning culture as characterized by:

- the willingness and competence to draw the right conclusions from the safety system; and
- the willingness to implement changes or reforms when necessary.

Learning involves ongoing reflection about current safety practices and beliefs, and the search for ways of eradicating or minimizing risks (Pidgeon 1998, 1991). Wiegmann et al. (2004) suggest that an effective incident-reporting system is the keystone in identifying vulnerabilities associated with safety management processes before safety incidents occur. However, an effective system improves safety only if an organization is willing to learn proactively and to adapt its operations. Thus, it is critical that managers respond to incidents (including near misses) and address identified health and safety issues in a timely manner. If workers observe that their reporting of incidents or deviations does not lead to any action, they will revert to seeing them as part of normal work process (Hale 2003) and organizations will lose valuable opportunities for learning and improvement. Previous studies of incident-reporting behaviour have identified the most frequent reasons for workers failing to report near miss incidents were that they 'were just part of a day's work' or 'nothing would get done' (Clarke 1998).

Learning is also associated with maintaining a questioning attitude. Hale (2003) argues that it is important for workers to have 'creative mistrust' in the risk control system. This means they are always expecting new problems, or new implications from old ones, and never believe their organizational culture or health and safety performance is ideal. But, nurturing a culture of creative mistrust also means there are explicit and supportive provisions for whistleblowers to inform management about latent safety problems. Hale (2003) argues causes for incidents and opportunities for improvements should be sought in the interaction of many causal factors rather than in individual behaviour. Therefore, solutions and ideas for health and safety improvement should be sought in many places and from many people, most notably frontline workers who work directly with the technology and the hazards.

There has been considerable research on the characteristics of a learning organization. These characteristics are relevant to learning about health and safety. They include:

- striving for continuous improvement and new ideas;
- ensuring all the individuals and teams are aware of the benefits of improving safety;
- learning from one's own experience and from the experience of others;
- sharing ideas and information internally and externally, and being open to and encouraging innovation;
- being mindful that things can go wrong and tolerating (but learning from) legitimate mistakes;
- allowing flexibility in searching for safer ways of working;
- actively learning from errors and failures rather than seeking to blame and/or find a scapegoat;
- questioning commonly held assumptions about what is safe and working to uncover latent (hidden) hazards in work systems; and
- fostering knowledge sharing throughout the organization (and the inter-organizational landscape), and crossing boundaries of teams, disciplines, and divisions.

The latter point is particularly pertinent in multidisciplinary, multi-organizational construction project environments.

Case Example 5.1 Learning Processes at a Large Infrastructure Construction Project

A learning initiative was implemented in a large infrastructure construction programme in Melbourne, Victoria. The multibillion dollar project took place over a five-year period and included construction of new rail track, new platforms at existing inner city train stations, new stations, a major upgrade to an existing suburban station, removal of two level crossings, 13 road and rail grade separations, and a new rail bridge.

The programme of construction work was delivered in six separate packages of work. The works packages were delivered using different commercial frameworks and contracting arrangements. Two packages were project alliances, while the others were delivered using a design and construct mechanism.

As the programme safety director explains, from the beginning of the work, the client sought to drive the message that 'there's no IP [intellectual property] with regard to safety'.

The client established a joint coordination committee (JCC) on which senior management from each contractor working on the programme and the rail operator were represented. The JCC operated as a governance committee, and was 'designed to look at key areas of the programme-wide performance where we could leverage benefits, synergies, be collaborative and so on'.

Under the JCC, discipline-based subcommittees were established, including a safety subcommittee. This 'created a forum where everyone met together, and everyone shared ideas, experiences, lessons learned'.

Case Example 5.1 (Continued)

The safety subcommittee met monthly throughout the programme of work. As the client's safety director explains:

> The safety manager from each package of work came in and we had a roundtable discussion about what was happening on each package. We shared incident information, trend information, and we also started to share initiatives. So if one package was running a sun smart initiative, we wouldn't say to the other five packages, "Run your own initiative." We'd share and use that initiative. We also started to develop knowledge management papers. So we've had knowledge management papers on issues such as underground service detection, management and permit systems, and plant/pedestrian separation, and we kept updating those knowledge management papers throughout the project.

The safety subcommittee held programme-wide safety events attended by representatives of all of the works packages. These events were primarily to enable works packages to share ideas and experiences of good practices and to learn from one another.

Sharing information provided the opportunity to capture important information that could be used to improve health and safety in subsequent phases of the programme of work. As the safety director explains, sharing health and safety knowledge

> … put us in a situation where, say, one contractor was trying a particular piece of technology… [they] were willing to share what they were doing and whether it worked or not, what the issues were with other contractors. And they were also willing to share with regard to past initiatives, whether they were successful or not, what they would do differently again. We essentially became the keeper of those knowledge management papers, the keeper of those initiatives, the keeper of the information that was being shared. And when I talk about "pay it forward", you're paying it forward to the contractors you're currently working with. But then, because you become the knowledge management repository, you're paying it forward to the next group of contractors.

These lessons were captured at the end of the programme of work and are now being used to inform the health and safety activities in a new programme of rail infrastructure construction work. In this new programme of work, the client is requiring successful tenderers to implement and evaluate new technologies that improve health and safety.

As the safety director explains:

> We want to experiment with technology and we want to share the learnings. We want the contractors to bid and propose to trial a new piece of technology during the delivery. And they need to write a report which we'll share with all our contractors. We don't want to have seven or eight different contractors try the same thing because they won't share the information between themselves.

5.8.7 Component 7: Trust in People and Systems

Hale (2000) distinguishes cultural influences from the management structures or systems established to deliver health and safety in an organization. For Hale, structures are important for maintaining good performance. Structures include elements of health and safety management systems such as management plans, policies, and procedures as well as performance monitoring and reporting mechanisms. However, there is a difference between the existence of these structures and the trust people put in them.

Trust is defined as an individual's tendency to rely on other people or structures in a risk situation. In relation to health and safety, trust is described as individuals' attitudes to, and expectations of, other people and the systems embedded within their organizational environments (Jeffcott et al. 2006).

Burns et al. (2006) describe how workers in a UK gas plant reported high levels of trust in their workmates, lower levels of trust in their supervisors, and even lower levels of trust in plant managers. These findings highlight the importance of understanding the expression of trust at different levels within an organizational hierarchy. These differences may be particularly acute in a hierarchical system of multilevel subcontracting, such as exists in construction.

Kines et al. (2011) identified 'trust in the general efficacy of the safety system' as an important aspect of organizational culture. The efficacy of the safety system is described as the system's ability to achieve safety objectives and goals; for example, the effectiveness of safety activities in reducing the number of accidents and injuries.

A recent study found that internal consistency is important in developing workers' trust in the way safety is managed (Conchie et al. 2011). Thus, for a safety system to be seen as trustworthy, the processes and practices defined by the safety system should align with the health and safety values espoused by the organization. It is important to ensure consistency between 'what is said' by the system and 'what is done' in practice (Simoms 2002).

Wiegmann et al. (2004) identify the quality and effectiveness of an organization's reporting system as being critical for health and safety effectiveness. However, Reason (1997) argues the most important determinant of reporting is trust. Trust-rich environments characterized by open communication are conducive to workers' willingness to identify and report abnormal events and errors (Jeffcott et al. 2006). The presence of a just culture is important in developing trust. Organizations with a just culture encourage and reward individuals who report safety-related issues, which enable the identification and resolution of latent error conditions in organizational systems. However, it is equally unacceptable to exempt from discipline unreasonable, reckless, negligent or malevolent behaviour that creates hazards or causes incidents. Thus, a just culture draws a clear line between acceptable and unacceptable actions.

Workers' trust in a safety system is also determined by the quality of information the system provides. Conchie and Burns (2009) investigated workers trust in a variety of information sources in the UK construction industry. They reported workers' trust in an information source is largely determined by the belief that the source's information is accurate. Workers reported a higher level of trust in the regulator and safety managers than they did in project managers and supervisors, perhaps reflecting their differing emphasis on production relative to safety.

Normally, it is assumed that trust in safety management systems is associated with positive safety outcomes and distrust is associated with negative safety outcomes. However, this is challenged by recent studies showing complete trust is actually undesirable. Jeffcott et al. (2006) reported rule-based trust (that is, a high level of trust in a system of rules) may have negative effects on safety, partly because it reduces flexibility to cope with abnormal situations not covered by pre-specified rules and procedures. Based on interview data, Conchie and Donald (2008) argue that both safety-specific trust and safety-specific distrust can have positive and negative functions in safety. Specifically, trust results in positive outcomes, such as open communication, reduced perceptions of risk among employees and improved employee confidence in safety management. However, complete trust may result in problems such as increased risk of mistakes and reduced personal responsibility for safety. These problems may be avoided by encouraging a certain level of distrust in the behaviour of others or in the safety system. This distrust finds expression in the form of questioning, monitoring, and checking. This means an enabling organizational culture needs both elements of moderate trust and moderate distrust.

5.8.8 Component 8: Resilience

Resilience has been defined as:

> ... the intrinsic ability of a system to adjust its functioning prior to, during, or following changes and disturbances so that it can sustain required operations under both expected and unexpected conditions.
>
> *(Hollnagel 2011, p. xxxvi)*

Hollnagel (2010) suggests resilience is related to four essential qualities or abilities. An organization should have the ability to:

1. **Respond to new or unusual situations in an appropriate way**
   This involves recognizing it is not enough to rely entirely on a set of policies and procedures because actual situations often differ from expected situations. This may be especially true in non-routine work. When irregular threats to health and safety arise, people need to respond in a way that ensures their health and safety in the new and unexpected situation. This requires adaptive behaviour and flexibility, but also a clear understanding about the boundaries of acceptable behaviour.
2. **Flexibly monitor what is going on, including its own performance**
   Flexibility means monitoring systems are assessed from time to time so they do not become normalized by routine practice. Monitoring enables an organization to deal proactively with matters that, if left unattended, may become critical in the near future.
3. **Anticipate future events that could impact on health and safety**
   This refers to the ability to go beyond the current situation, and to anticipate what may happen in the future. Anticipation enables an organization to pre-empt and deal with potential problems and new situations as they arise.
4. **Learn from experience**
   The ability to learn from what has happened by making changes to procedures, roles, and functions, or even to the organization itself. This learning ability enables the organization to deal with dynamic and complex environments.

Reason (1998) argues organizations should have an abiding concern with failure and recognize that their safety systems are fallible. A belief that safety systems are infallible can make people 'forget to be afraid'. Thus, a resilient organization knows hazards are never completely eradicated and that errors, unexpected situations, and incidents are inevitable. Unexpected adverse events are seen as important indicators of areas in which the safety of a system can be improved (Olive et al. 2006).

An organization's resilience is reflected by flexibility and variability in operations. Many organizations attempt to reduce the number of unsafe acts by requiring employees to comply rigidly with procedures. They see errors and violations as workers' deviations from standard procedures and subject to sanctions and disciplines. Unfortunately, focusing on punishment leads to the organization losing opportunities to reflect on current procedures and analyse systemic causes of workers' unsafe acts. Collective mindfulness is claimed as an essential component of organizational resilience (Weick et al. 1999). According to Weick et al. (1999), collective mindfulness is the result of a number of cognitive elements, including preoccupation with failure, reluctance to simplify interpretations, sensitivity to operations, commitment to resilience, and under-specification of structures.

For Reason (2000), collectively mindful organizations are characterized by:

- working hard to extract the most value from the little data they have about rare events and catastrophic failures;
- being active in creating a reporting culture that encourages or rewards people who report incidents and near misses;
- working on the assumption that what seems to be an isolated failure may stem from a number of 'upstream' causal chains – they strive for system reforms rather than applying local repairs; and
- being aware that system failures can take a variety of yet-to-be-encountered forms – looking out for unexpected paths through which active failures or latent conditions can defeat system defences.

5.8.9 Component 9: Engagement

Employee engagement is defined as:

> Personnel from all levels of the organisation are involved in decision making, safety planning and providing ideas for improvement. Employee participation and feedback are actively sought.
>
> *(Health and Safety Executive 2005b)*

Workers' participation and involvement in workplace health and safety activities is linked to reduced incidents and injuries (Neal and Griffin 2006; Christian, et al. 2009). In some instances, this manifests in empowering workers to use their judgement and knowledge to develop safe and healthy work practices.

Kines et al. (2011) suggest delegation of power demonstrates that managers trust workers' ability and judgement, and value workers' ideas about improvements that can be made to work processes. Workers who feel empowered tend to:

- have higher motivation to 'make a difference';
- go beyond normal duties to secure organizational safety; and
- take more responsibility for ensuring safe operations (Wiegmann et al. 2004).

Research has identified leader behaviours that are influential in engaging employees in safety participation. Clarke and Ward (2006) found workers are more likely to participate in health and safety activities when managers share health and safety information, and actively seek to involve workers in strategic health and safety-related decision making. Supervisors play a particularly important role in engaging frontline workers by communicating that they value workers' ideas and trust their judgements about working safely. In Chapter 8 we will return to the topic of engagement when we describe using participatory video to elicit workers' ideas for redesigning work processes.

5.9   The Organizational Culture Maturity Continuum

It is recognized that organizational cultures progress through different stages of maturity. Hudson (2007) suggests merely defining and describing components of an organizational culture that can enable health and safety will not help organizations develop such cultures. He advocates understanding culture using an evolutionary model in which organizations are placed on a continuum from those at an advanced stage of cultural development to those at a less advanced stage. It is argued that defining intermediate stages can assist organizations to engage in culture change in manageable steps.

Hudson (2007) developed a five-level framework for describing the progressive development of a culture that supports safety. These levels are shown below.

1. Pathological: Who cares about safety as long as we are not caught?
2. Reactive: Safety is important: we do a lot every time we have an accident.
3. Calculative: We have systems in place to manage all hazards.
4. Proactive: We try to anticipate safety problems before they arise.
5. Generative: Work health and safety is how we do business around here.

This framework emerged from interviews with senior managers in the oil and gas industry. They identified aspects of the organization they believed were important elements of a safety culture in the industry. Interviewees were asked to describe how an oil company would function in relation to each element at each of the five levels of cultural maturity (that is, from pathological to generative). Parker et al. (2006) used these five levels to develop a framework that can be used by organizations in the oil and gas industry to understand their organizational cultures and safety impacts.

A variation of Hudson's five-level culture framework was developed for the UK healthcare sector. Ashcroft et al. (2005) report on the feasibility and face validity of a five-level healthcare culture maturity model. More recently, the five levels specified by Hudson, Parker, and others were used to develop an organizational culture maturity assessment tool for analysing the health and safety implications of culture in the oil and gas industry in Brazil (Filho et al. 2010). Ayers et al. (2013) also used Hudson's model to analyse cultural maturity in the way construction companies engage in consultation with workers about health and safety.

A culture maturity continuum was developed for the Australian Constructors Association (ACA). The continuum was based on the nine components of organizational culture identified as being relevant to work health and safety, and using the five levels specified by Hudson. For each of the nine components, descriptors were developed to reflect the five levels of maturity.

The resulting Organisational Culture Maturity Continuum was then subject to validation and testing in four workshops and a series of interviews. In total, 65 industry representatives participated in the workshops, including senior managers from construction organizations, health and safety managers, trade union representatives, and other managers and professionals.

In the workshops, participants used the maturity model to assess a fictitious organization described in a scenario. However, this was an oversimplified description of an organization that could not reflect the true complexity of real-life organizational environments. Participants noted that using the maturity model in a real construction organization would present challenges for people in making realistic and reasonable assessments of their own and others' levels of cultural maturity.

Based on their reading of the scenario, readers found it easy to understand and apply the descriptors associated with each component. Although there was some variation between participants in positioning the fictitious organization on the maturity continuum, the majority of workshop participants acknowledged the descriptors as presenting a coherent set of guiding statements that could be used to interpret the nine components.

The workshop participants generally understood the components (including their associated descriptors) as existing along a continuum. However, it was noted that the descriptors enabled discernment of an 'overriding impression' of organizational maturity, as distinct from considering an organization as discretely fitting within one level of maturity or another. Participants' discussions of the variance in assessments did not reveal dissatisfaction with the descriptors or levels, but indicated an appreciation that any such assessment is inherently subjective, and different people may have different points of view.

Workshop participants noted that by combining the components and the descriptors of each of the five levels of maturity, the model stimulated a discussion about what constitutes a mature organizational culture. They commented that the model promoted deeper consideration about how some managerial behaviours can influence health and safety and, as a consequence, they were better equipped to understand organizational behaviours and the messages they send from different viewpoints.

Participants commented that the model could be used to prompt conversations within organizations about managerial behaviour and organizational priorities. Participants acknowledged the importance of being able to review an organization (or its component parts) and suggested the maturity model would be a useful tool to focus discussion about organizational and managerial behaviours that can impact health and safety.

Some participants were familiar with the words originally used by Hudson to describe the five levels of maturity – that is, pathological, reactive, calculative, proactive, and generative. However, several participants expressed the view that these words were too abstract and not in common use. It was perceived that using these words as terms for the levels of cultural maturity could render the meaning difficult to comprehend. The maturity continuum was revised on the basis of this feedback. A five-level framework is still used. However, the framework now reflects participants' comments that cultural maturity development is best understood as a continuous progression along a continuum. In response to that understanding, it was deemed appropriate to provide verbal 'anchors' for desirable and undesirable levels of maturity, but to omit labels for each of the levels in between.

This decision reflects the observation, made by many participants in the workshops, that it is difficult to position an organization in a discrete cultural maturity level – in many cases they fall somewhere between two levels. A cultural maturity continuum or spectrum was considered preferable. The verbal anchors reflecting high and low levels of cultural maturity are now 'Enabling' and 'Impeding'. These anchors also reflect the understanding of health and safety as an outcome of the broader organizational culture that can either impede or enable health and safety in a workplace.

The revised Organisational Culture Maturity Continuum, which was further developed and expanded by the ACA with input from member companies, is presented in Table 5.7.

5.10   Conclusions

The term 'safety culture' is used widely and, in many instances, is not clearly defined. Sometimes the safety culture of an organization is treated as a thing that an organization either has or it does not have. It is assumed that organizations that have a safety culture will perform well in workplace safety (though the potential cultural influences on workers' health are not usually mentioned). Presumably organizations that perform poorly in WHS do not have a safety culture.

Positioning the safety culture as being distinct from the broader organizational culture creates an artificial segregation between WHS and the assumptions, values, and norms that influence a wide range of behaviours. These behaviours, for example in relation to communication, worker engagement, organizational learning, or establishing responsibility, authority, and accountability, are not necessarily WHS-focused, yet they all have the potential to impact workers' WHS. Thus, it is potentially more helpful to consider the way in which broader organizational (and project) cultures impact WHS.

Attempts to impose a 'safety culture' from the top down are commonplace but fraught with difficulty because organizations are complicated and multicultural. Subcultures form at a local level within organizations as social groups develop shared meanings of policies, practices, and events. Safety culture programmes designed to create uniformity around a dominant (often managerial) culture can reduce trust and may also inadvertently disturb locally good WHS practices. Thus, it is very important that any culture change initiatives are carefully designed and do not oversimplify the organizational environments in which they are to be implemented.

Table 5.7 The organizational culture maturity continuum.

Leadership

Scope: The ability of managers at all levels to promote transformational thinking and change through positive engagement and actions with the workforce.

Impeding ⟵⟶ Enabling

Impeding				Enabling
Managers are more concerned with operational issues than health and safety matters.		Managers seek to manage health and safety matters to avoid prosecution rather than protect workers/contractors.	Managers seek information about health and safety matters, performance, and incidents to help manage improvement.	Managers at all levels actively integrate health and safety matters into business operations, and participate in and act on conversations and improvement plans.
Managers respond negatively to all feedback about health and safety.	Managers are interested in health and safety when something goes wrong, but do not follow-up on actions.	Managers attend formal health and safety activities to simply meet their required quota.	Managers are visible in the workplace and demonstrate active interest in the health and safety of people.	Managers demonstrate genuine concern for people and a desire for continual improvement in all health and safety matters.
Managers and workers/contractors are suspicious of each other and don't talk about health and safety matters.	Managers occasionally talk to workers/contractors on some health and safety matters, but don't seek their opinions.	Managers discourage health and safety reporting by actively blaming workers/contractors when things go wrong.	Managers actively converse with workers/contractors on health and safety matters, and listen to concerns.	Managers consistently involve themselves in health and safety matters and improvements, and respond to concerns.
Managers only involve themselves in health and safety messaging to promote their position or self-interests.	Managers change their messaging based on health and safety circumstances.	Managers sometimes involve themselves in health and safety to understand if matters may affect them.	Managers are actively involved in health and safety matters and messaging to better understand matters and sometimes support outcomes.	Managers encourage open, blame-free reporting by workers/contractors on all health and safety matters to encourage learning and support continual improvement.

Organizational goals and values

Scope: The organizational attitude towards the role of health and safety in its operations, and its place within the priorities of the organization.

Impeding ⟵⟶ Enabling

Impeding				Enabling
Managers consistently prioritize cost minimization at the expense of health and safety.	Health and safety expenditure is minimal and considered only when incidents and/or client pressure is applied.	Managers make public statements about the importance of health and safety, but expenditure on health and safety is regarded as discretionary.	Health and safety and profitability are juggled, but some project delays and costs are borne to improve health and safety.	The organization invests in innovation to find ways to make work safer and healthier.
Health and safety is seen as a cost to the organization and an impediment to production.	Health and safety issues only become relevant if they affect project schedule and production.	There is an understanding that minimum health and safety standards must be maintained so that production is not affected.	Health and safety resources are regarded as important to the business and can influence business decisions to improve production.	Health and safety is an integrated component of the organization's strategy, business activity, and decision making.
Profitability is the only concern of managers.	Health and safety is regarded as a bureaucratic impediment to work and profitability.	Health and safety and profitability are juggled (as opposed to being balanced).	Health and safety is regarded as important because it is recognized it can contribute to financial success.	Health and safety is seen as able to contribute to profitability.

(Continued)

Table 5.7 (Continued)

Communication

Scope: How an organization consults and communicates in the delivery of health and safety messages.

Impeding ⟸　　　⟹ Enabling

Impeding				Enabling
Health and safety information is not communicated to workers/contractors.	Limited and intermittent health and safety information is communicated to workers/contractors.	Managers share limited health and safety information with workers/contractors.	Health and safety information is routinely and regularly communicated to workers/contractors.	The organization actively and openly shares health and safety information with workers/contractors.
Communication is one way and directive.	Ad hoc communications and generic slogans are visible but do not match workplace management values, and any positive impact associated with these soon diminishes.	Health and safety information is provided on an 'as needs' basis.	Two-way communication is actively encouraged.	Strategic health and safety information is openly shared.
Conflicting messages about the importance of health and safety are conveyed.	Safety messages, when given, are sometimes unclear.	Communication tends to focus on day-to-day operational issues.	Suggestions and ideas provided by workers/contractors regarding health and safety improvements are taken seriously and implemented where possible.	Health and safety communication is frequent and effective.
Communication is mainly top-down, usually occurring to resolve an issue.	There is little or no opportunity for bottom-up communication from workers/contractors to management about health and safety concerns.	Workers/contractors communicate their health and safety concerns and ideas to managers, but their suggestions for improvements have little impact.	Health and safety communication is a strong and consistent two-way process.	Managers receive as much health and safety information as they give, and act on the information they receive.

Scope: How corporate structure supports on-site culture surrounding health and safety.

Impeding				Enabling
Work is designed and scheduled in a way that creates excessive time pressure, workload, stress, and fatigue.	Managers and workers deal with stress and workload problems as they arise.	An effort is made to improve workers' health and wellbeing, but work schedules still demand excessive hours.	Work is restructured so far as possible to support health, wellbeing, and work–life balance.	Jobs and work conditions are specifically designed to positively promote health, wellbeing, and work–life balance.
Obstructive and uncooperative relationships exist between groups and functional areas, such as health and safety and project management teams.	There are low levels of cooperation and poor information flows between work groups and functional areas.	Cooperation and communication between work groups and functional areas is sufficient to get work done.	Work groups and functional teams work hard at sharing information and cooperating to improve workers' health, safety, and wellbeing.	Cross functional cooperation and team work are effective and focused on finding ways to support workers in working healthily and productively.
People feel overwhelmed and unable to perform work in a healthy and safe manner.	Health and safety are treated as an individual's responsibility.	Management and workers are provided with basic knowledge, skills, and competency in health and safety.	Workers at all levels have the knowledge, skills, and ability to work in a healthy and safe way.	People are empowered to resolve health and safety issues, and feel confident reporting errors and violations so that better systems of work can be designed.
No effort is applied to managing the hazards in the physical environment (for example, no housekeeping, no guarding).	The physical workplace, amenities, and equipment reflect minimum standards and improvements made only when externally influenced.	The physical workplace, amenities, and equipment are at basic industry standard.	The physical workplace, equipment, and facilities reflect above-average industry practice.	The organization actively invests and experiments with ways to provide a healthy and safe work environment for all workers/contractors.

(Continued)

Table 5.7 (Continued)

Impeding ⟵			⟶ Enabling
Organizational structure does not include health and safety support or integration.	Organizational structure supports health and safety compliance to minimum requirements.	Health and safety is considered in project design and planning prior to and during construction activities.	Organizational structure supports health and safety innovation, information sharing, and change management.
Information and knowledge are seen as power, and withheld within the organization.	Information and knowledge sharing is based on compliance and protecting the organization from litigation.	Health and safety information and knowledge is readily shared.	Health and safety information and knowledge is used to improve systems of work across the organization.

Scope: How managers, workers, and contractors view their responsibility, authority, and accountability surrounding health and safety within an organization.

Responsibility, authority, and accountability

Impeding ⟵			⟶ Enabling
At all levels of the organization – management, workers, and contractors – they believe health and safety is someone else's responsibility.	Health and safety responsibilities and accountabilities are poorly communicated and understood, change frequently depending on circumstance, and outcomes are uncertain when held to account.	People think it is the job of the health and safety professionals to 'police' the workplace.	Health and safety is treated as everyone's responsibility.
			At every level, there is a willingness to take personal responsibility for health and safety.
Health and safety responsibilities and accountabilities are not communicated and understood.	People only think about their health and safety responsibility when things go wrong.	Unsafe practices are sometimes reported, but personal responsibility is avoided.	Line managers take responsibility for health and safety in their work areas, and the role of health and safety professionals is understood as one that provides technical input.
			When incidents and issues arise, managers look inwards as well as outwards to identify causes.
			When people work in ways that are unacceptable, the organization treats them in a

Impeding				Enabling
People turn a blind eye if they observe an unsafe practice, or do not report for fear of retribution.	People are concerned about health and safety, but do not intervene when they see something wrong. People are never held to account for their health and safety responsibilities.	People are not equally held to account to their health and safety responsibilities. People rarely think of their moral responsibility towards health and safety.	People actively stop unsafe practices when they are observed.	All personnel actively demonstrate care and concern in looking after both their health and safety and that of others. Positive health and safety behaviours are driven by strongly held collective norms and expectations. People feel confident reporting errors and violations so that better systems of work can be designed.
People do not feel they have authority to act in a way that is equal with their role and responsibilities.	People feel that they have the authority to act only when a breach of the law is being committed.	People do not feel they have total authority to act in a way that is equal with their role and responsibilities.	People will stop work when encouraged by managers. Managers reward people who stop work.	People have no hesitation in stopping work practices if they have health and safety concerns. Managers actively support people who stop work.
Health and safety responsibility and accountability are avoided for fear of being blamed when things go wrong. When incidents happen, it's the injured person who is held responsible for fault.	When incidents happen, people look to assign personal blame.	When incidents happen, investigations focus on identifying and rectifying immediate causes, and human error is often the focus.	When incidents happen, investigators consider organizational factors that contribute to human errors.	There is a strong understanding that people who undertake work have a right to contribute to the design of work.

(Continued)

Table 5.7 (Continued)

Learning

Scope: How lessons from health and safety incidents are utilized, actioned, and communicated in an organization.

Impeding ⟷ Enabling

Impeding				Enabling
Health and safety performance data is not systematically collected and analysed.	Health and safety performance is measured using only the incidence of serious/reportable injury (that is, lost-time injury).	Health and safety performance is measured using 'lagging' indicators, such as the occurrence of incidents, injuries, and illnesses.	Health and safety performance is measured using mainly lagging indicators, but some positive (lead) indicators are also used.	Health and safety performance is measured using a balanced mix of lagging indicators and positive performance (lead) indicators.
The causes of incidents, errors, and deviations from procedures are not analysed.	Incident investigations focus on identifying immediate causes. No attempt is made to identify the systemic causes of incidents.	Incident investigations consider broader workplace conditions and work processes as possible causes.	Incident investigations attempt to identify systemic causes of incidents, including those relating to organizational culture, risk management processes, design of projects, and project management practices.	Incident investigations are rigorous and focused on uncovering systemic causes of incidents.
The analysis of incidents, errors, or deviations from procedures focuses on identifying someone to blame.	Recommended preventive actions are mainly 'behavioural'.	Recommended preventive actions address workplace and work process improvements.	Recommended preventive actions address organizational issues.	Recommended preventive actions address 'upstream' issues, including safety in design and project planning.
Feedback is not sought from workers/contractors and others about the effectiveness of health and safety policies and processes.	Feedback is sought from workers/contractors and others about the effectiveness of health and safety policies and processes, but feedback is never acted upon.	Feedback is sought from workers/contractors and others about the effectiveness of health and safety policies and processes.	Feedback from workers/contractors and others about the effectiveness of health and safety policies and processes is sought and informally used to inform health and safety improvement actions.	Feedback from workers/contractors and others about the effectiveness of health and safety policies and processes is systematically analysed and considered in formal health and safety planning processes.

Impeding				Enabling
Health and safety training provides basic minimum requirements only.	Health and safety training is generic and compliance-focused.	The organization provides structured health and safety training programmes to workers/contractors and stakeholders.	Health and safety training is reflective and allows for intelligent application.	Health and safety training is engaging, relevant, and effective in transferring knowledge to workers.
No actions are proposed for ongoing health and safety improvement.	There is no attempt to transfer health and safety lessons from project to project.	Health and safety improvement is usually driven by outcomes after a serious incident has occurred.	Post-project reviews capture valuable health and safety information that is carried forward to improve performance in subsequent projects.	Workers'/contractors' perceptions and views of the organization's health and safety processes and performance are actively sought.

Resilience

Scope: The ability of an organization to adapt to change and promote innovative practices that lead health and safety.

Impeding				Enabling
Health and safety policies and procedures are rigid and cover most eventualities.	Health and safety policies and procedures comply with minimum legislative requirements.	Health and safety policies and procedures are developed to prevent incidents from occurring.	Health and safety policies and procedures comprehensively cover the organization's activities and some opportunity exists for change.	Health and safety policies and procedures are open to continual improvement and a process exists for consultation, review, and improvement.
Managers endorse health and safety policies and procedures as a failsafe way to avoid incidents.	Managers give no consideration to whether health and safety policies and procedures can be complied with.	Managers regulate intended health and safety behaviours through policies and procedures that relate to known hazards/risks.	Managers sometimes consult on health and safety policies and procedures, and they are extensively integrated into training provided to workers/contractors.	Managers actively consult with workers/contractors, and seek feedback and changes on health and safety policies and procedures to ensure they remain applicable.

(Continued)

Table 5.7 (Continued)

Impeding ⟵ ⟶ Enabling

Impeding				Enabling
Health and safety policies and procedures only exist to respond to litigation and protect managers.	Health and safety policies and procedures are written to prevent the last incidents that happened from recurring.	The number of health and safety policies and procedures keeps growing in response to incidents and identified hazards.	Health and safety policies and procedures exist for the purpose of promoting good practices across the organization	Health and safety supports creative thinking about risk management to encourage leading practices.
The success of health and safety policies and procedures relies solely on worker behaviours.	Health and safety policies and procedures can be achieved only if workers'/contractors' behaviours are strictly controlled.	Minimum standards for health and safety policies and procedures are dictated to workers/contractors to follow, regardless of their practicality and ability to be implemented.	Health and safety policies and procedures have some worker/contractor input, and some flexibility within the boundaries of acceptable practices.	Leading practices are actively endorsed in health and safety policies and procedures. Engineering controls or better are sought for high risk activities to protect workers/contractors against inadvertent or unintended behaviours.

Engagement

Workers/contractors are not engaged in organizational or project-level health and safety activities.	Workers/contractors are invited to participate in health and safety activities only after a serious incident has occurred.	Some workers/contractors are involved in health and safety-related activities. 'Carrot and stick' reward and punishment are used to influence engagement.	Workers/contractors are generally encouraged to participate in the organization's health and safety activities. Engagement programmes are utilized to influence workforce engagement in health and safety.	All workers/contractors have input into decision making as it relates to health and safety.

Scope: How managers engage workers/contractors in health and safety matters and the influence and outcomes that arise from these engagements.

Impeding				Enabling
Managers have no interest in engaging workers/contractors in health and safety activities.	Managers will only ask for worker/contractor input into health and safety activities when required to do so.	Formal consultation mechanisms are in place but not fully embraced or understood by management and/or workers/contractors.	Managers actively seek input from workers/contractors relating to operational aspects of health and safety, including work planning and the development of procedures/rules.	Workers/contractors feel they are able to influence health and safety activities in the organization/project.
Workers/contractors are not asked to provide health and safety input.	Minimal effort is put into consultation activities. Worker/contractor opinions are often dismissed.	Workers/contractors are asked to provide input on basic health and safety issues like training, safety equipment, and housekeeping.	Workers/contractors are regularly consulted on health and safety as standard practice. Health and safety issues raised are acted on and feedback provided.	Managers actively seek input from workers/contractors concerning strategic aspects of health and safety in the organization/project, including issues of work design and the operation of the health and safety management system. Management is visible in the workplace and seeks information from workers on how to improve health and safety.
Work procedures/rules/processes are imposed and mandated by the client.	Management reacts to worker/contractor poor health and safety performance.	Management holds regular worker/contractor reviews and discusses health and safety improvements. Workers/contractors are included in health and safety meetings and have input into basic health and safety issues.	Management programmes include formal worker/contractor engagement forums. Workers/contractors health and safety issues are acted on and feedback provided.	Active health and safety engagement and participation is the norm.

(Continued)

Table 5.7 (Continued)

Trust in people and systems

Scope: How comfortable workers and contractors are with the health and safety reporting system in place in an organization, and their confidence and trust in the system achieving health and safety improvements.

Impeding ⟵⟶ Enabling

Impeding				Enabling
Health and safety systems are designed and implemented solely to protect the company and its profits.	Health and safety systems are compliance-focused and creating a paper trail seems to be the most important outcome.	Health and safety processes and initiatives are meaningful and workers/contractors perceive them to be well-motivated and beneficial.		There is systematic follow-up to ensure that newly implemented health and safety initiatives are having the desired effect.
Health and safety systems are unstructured and poorly documented.	The health and safety system is never reviewed or evaluated – even when multiple incidents happen.	Auditing inside and outside the organization provides an opportunity to review and improve the quality and effectiveness of organizational health and safety activities.		Innovative solutions to identified health and safety challenges are pursued, implemented, and rigorously evaluated.
There is no reporting culture.	Workers/Contractors do not report health and safety issues because they believe nothing will be done to resolve them.	Workers/Contractors feel uncomfortable and are reluctant to report health and safety issues	Workers/Contractors are somewhat uncomfortable reporting errors or deviations from procedures, but are willing to do so because they hope that this will result in health and safety improvement.	Workers/Contractors feel very comfortable reporting errors or deviations from procedures and firmly believe that this will result in health and safety improvements.
Incidents are denied and investigations are undertaken in secrecy.	Investigations identify who is to blame after an incident and prevention strategies focus on behavioural control.	Incident investigation collects a lot of data and produces lots of action items, but opportunities to address real issues are often missed.	Most incidents, errors, and deviations from procedures are reported and investigated.	Incident investigations are open, transparent, and search for a deep level of understanding of how incidents happen.

Discussion and Review Questions

1 Should workplace health and safety be viewed as a component or outcome of the organizational culture? What are the implications of adopting these differing points of view?

2 Can culture be effectively imposed from the top down in an organization or project?

3 What aspects of an organizational or project culture enable and/or impede workplace health and safety?

4 How do organizational or project cultures vary in their level of maturity? Do cultures change over time? What might be the catalyst for change in a construction organization?

Acknowledgements

Research presented in this chapter was funded by the Australian Constructors Association (ACA) and first appeared in a report titled 'Health and Safety Culture'. The Organisational Culture Maturity Continuum was originally developed by the research team and further refined and improved by the ACA.

6

Understanding and Applying Health and Safety Metrics

Helen Lingard, Rita Peihua Zhang, Payam Pirzadeh, and Nick Blismas

School of Property, Construction and Project Management, RMIT University, Melbourne, Victoria, Australia

6.1   The Measurement Problem

Construction organizations routinely measure their workplace health and safety performance, using a variety of different measurement methods and metrics. In some cases, measurement is undertaken in response to contractual or legislative reporting requirements. In other instances, measurement is used to monitor, benchmark, or improve performance. Appropriate measurement of health and safety outcomes is less frequently used to evaluate the effectiveness of a health and safety intervention, although there have been calls for more rigorous evaluation studies (Robson et al. 2001).

The way health and safety is measured in construction organizations and projects has been questioned. In particular, the usefulness and validity of so-called lag indicators, including injury frequency rates, have been challenged. There has been a shift in emphasis towards measures of system safety that are expected to *lead* changes in the incident rate. These measures are sometimes based on the frequency or quality of health and safety management activities, and sometimes based on workers' perceptions of the state of safety in the work environment. However, the validity of so-called lead indicators has not been rigorously evaluated, and the time-dependent relationships between expected lead indicators and injury or incident outcomes are unclear.

In this chapter we critically review commonly used health and safety performance metrics. We present analysis of data collected at a five-year construction project showing the complex relationships between lag and lead indicators over time. This analysis opens questions as to whether the terms 'lag' and 'lead' – which have been uncritically adopted from the economics and finance field – are appropriate to use for work health and safety (WHS). We consider the usefulness of measuring workers' perceptions of the work environment (the safety climate) and we identify some of the problems inherent in assuming climate is homogeneous and stable in construction organizations and projects. We identify an opportunity to develop alternative metrics and methods to capture the quality of workplace health and safety outcomes, particularly about safety in design. The chapter concludes with a discussion about how metrics are sometimes used to drive performance in construction projects, with a cautionary note about the

Integrating Work Health and Safety into Construction Project Management, First Edition.
Helen Lingard and Ron Wakefield.
© 2019 John Wiley & Sons Ltd. Published 2019 by John Wiley & Sons Ltd.

unintended consequences that can arise when metrics are linked to commercial arrangements.

6.2   Why Measure Work Health and Safety Performance?

The measurement of health and safety performance is undertaken to inform, support, and evaluate organizational health and safety management activities. The management of workplace health and safety relies on the 'systematic anticipation, monitoring and development of organisational performance' (Reiman and Pietkäinen 2012, p. 1993).

Regular measurement of health and safety performance enables detection and resolution of problems, and provides information needed to make proactive decisions and evaluate the effectiveness of initiatives. The use of performance data to identify and respond to changes in workplace health and safety conditions is a feature of resilient, high reliability organizations (Cooke and Rohleder 2006).

Most construction organizations engage in strategic, operational, and project-planning processes, which include establishing specific health and safety objectives. Key performance indicators – or measurable values – are then specified for health and safety performance to enable organizations to assess whether objectives are being met. Performance measurement can also be used to identify areas to target for improvement.

Although measuring workplace health and safety performance is routinely undertaken and considered useful, there is considerable disagreement about how best to measure the health and safety performance of a construction project. As we will show, the choice of indicators used, in conjunction with establishing health and safety goals, can influence management behaviour, sometimes producing unintended consequences.

6.3   Different Types of Performance Indicator

Different types of indicators are currently used to measure WHS performance in the construction industry. However, it is noteworthy that far greater attention is paid to measuring safety performance than to measuring health-related risks and impacts. We discussed the relative neglect of occupational health in Chapter 4.

Kjellén (2009, p. 486) defines safety performance indicators as 'the metric[s] used to measure the organisation's ability to control the risk of accidents'. Harms-Ringdahl (2009) defines safety indicators as 'observable measures that provide insights into a concept – safety – that is difficult to measure directly' (p. 482).

Decisions about which indicators should be used to measure an organization or project's health and safety performance are ultimately informed by one's understanding or beliefs about what constitutes and explains workplace health and/or safety (Reiman and Pietkäinen 2012).

The terms 'lag' and 'lead' have been applied to different types of performance indicator for workplace health and safety. These terms were borrowed from economic and financial modelling. In economics, a lead indicator is something that changes before the economy changes; for example, building permit approvals and stock prices (Wreathall 2009). However, as Kjellén (2009) argues, these terms were introduced to the field of workplace health and safety without full consideration of their meanings. For workplace health and safety, the lag/lead terminology implies a distinction between proactive measures of the state of workplace health and safety, and retrospective measures of past (mostly undesirable) health and safety outcomes. However, the dependencies and temporal relationships among so-called lag and lead indicators of health and safety are very unclear. Later in this chapter we will present data showing that it may be unjustified to assume that proactive measures of management activities will 'lead' changes in injury or incident rates. But first we will consider the relative advantages and disadvantages of using injury or incident rates to measure workplace health and safety performance.

6.4 Lag Indicators

Incident or injury frequency rates are the most frequently used lagging indicator of safety performance in the construction industry. There are standardized ways to calculate lost time injury frequency rates (LTIFRs) and total recordable injury frequency rates (TRIFRs). Such indicators are useful because they are:

- relatively easy to collect;
- easily understood;
- easy to use in benchmarking or comparative analyses; and
- useful in identifying trends over time (NOSHC 1999).

However, these measures have been criticised as being statistically meaningless and focusing too much attention on the absence of negatives rather than the presence of positives in relation to workplace health and safety (Dekker and Pitzer 2016).

Because recordable incidents and injuries have a statistically low probability of occurrence over short timeframes, they are usually neither valid, nor stable, when measured at a single construction project (Hopkins 2009a). Hopkins (2009b) terms this the 'zoom' effect, referring to the fact that, even in very large construction projects, the frequency of accidents/injuries is insufficient to calculate a meaningful rate. Even a stable safety system will produce a variable number of injuries/incidents (Stricoff 2000). In addition, the absence of injuries/incidents does not necessarily mean a workplace is safer than another workplace at which an injury/incident occurred in the same period (Cadieux et al. 2006).

But perhaps more fundamentally, incident/injury rates are retrospective indicators capturing things that have already gone wrong. They measure the absence, rather than the presence, of safety (Arezes and Miguel 2003) and therefore cannot be regarded as a direct measure of the level of safety in a work system (Lofquist 2010). Weick (1987) describes safety as a 'dynamic non-event' and argues that, by definition, non-events cannot be counted.

The reliance on incident rates as the method of monitoring safety performance can have serious consequences. For example, Lofquist (2010) describes how relying on incidents as a safety indicator resulted in the failure to recognize a marked deterioration in safety that occurred in the Norwegian civil aviation industry during a period of organizational change. Pilots and air traffic controllers had observed a gradual decline in safety standards, but because no incident had occurred, decision makers were unaware of the negative safety impact of the organizational change programme. Thus, a low incident/injury rate does not guarantee that safety risks are being controlled or that incidents/injuries will not occur in the future (Mengolinim and Debarberis 2008).

The use of injury/incident rates to underpin incentive schemes can also cause reporting problems. Tying incentives – such as management performance appraisals, bonus payments, or future tendering opportunities – to injury/incident rates can encourage underreporting (Cadieux et al. 2006; Sparer and Dennerlein 2013). Pedersen et al. (2012) describe how group-based rewards for periods of accident-free working can encourage underreporting. Research also shows that workers who perceive they have low levels of job security are less likely to report injuries and accidents (Probst et al. 2013). In fact, the greater the emphasis placed on injury/incident rates in commercial incentive schemes, the less useful these measures are likely to be, because people learn how to manipulate them (Hopkins 2009b). Research into the use of commercial frameworks to drive construction project health and safety performance in the Australian construction industry highlights the unintended consequences associated with focusing too heavily on measuring lost time injuries.

Case Example 6.1 (Un)reliability of Injury Frequency Rates

Clients often require contractors to report their health and safety performance and typically use lagging indicators as the main performance metric. Although seemingly objective, these indicators can be manipulated and, especially when health and safety performance is built into commercial arrangements, they may be subject to underreporting. For example, construction contractors interviewed in relation to infrastructure construction work described how, under some arrangements, 'if you have a Lost Time Injury (LTI) you're going to lose 25 per cent of your bonus or etcetera… and companies become fantastic at hiding it'. Another contractor explained:

> The problem with all of those metrics [Lost Time Injury Rates] is that they're manipulated… and that actually undermines everything that happens onsite… I personally have been put under pressure to manipulate data for their statistics because they've [referring to senior managers] got bonuses that relate to it.

Determining whether something is reportable as a LTI can be subject to manipulation. One contractor described how

> … if someone got injured there would always be a [management] person with them at the doctor ready to say: "this person has work capacity, please don't give them an unfit for work certificate. We'll find them work to do."

Another described how 'people game it, and don't come clean on incidents, and you get a cover up kind of mentality'. This behaviour was also acknowledged to occur by some client representatives. One of whom indicated that on a previous project:

> Essentially we drove the wrong behaviour. One of the measures was Lost Time Injury Frequency Rate, and it drove a behaviour where contractors were managing the stat, not the injury.

(Continued)

Case Example 6.1 (Continued)

One contractor representative also observed that counting lost time injuries may not reflect the quality of health and safety management effort in a project, describing an incident in which

> … we've got an LTI for a chap who was doing work under the safety management system. He bent over to pick up a conduit, hadn't actually picked up the conduit or done anything, and he strained his back. He could have been bending over to do up his shoelaces for example. Maybe the safety system could have done something different, but I don't believe so.

Clients also acknowledged that measuring health and safety by counting the frequency of lost time injuries is a very blunt approach that inadequately captures important aspects of health and safety performance. One client described how

> … every month we had a monthly report and there was a graph tracking – our key tracking mechanism, was the Lost Time Injury Frequency – and I mean we were always trying to go to zero and really pushing that all injuries are avoidable but, you know, we recognised that some injuries do occur.

Some clients expressed a preference for measuring cultural aspects of worksite health and safety. One commented:

> You could then measure both maturity, safety climate, engagement – that sort of stuff on a six monthly basis. Then I think we would start to see we will be rewarding the outcomes that we truly want rather than the ones that we go, "well that's easy to measure, therefore we'll incentivise it".

Contractors also expressed a preference for measuring health and safety performance using leading indicators. One explained that

> … we've got key performance indicators that are based on lead indicators, instead of lag indicators. So we're not looking back, we're looking forward. So like positive insights, positive investigations, leadership visits, things that are done from a positive perspective that could improve our safety behaviour and performance.

6.5 Alternative Indicators

The well-documented criticisms of injury and incident rates as a measure of workplace safety performance have led to development of different ways to quantify the state of safety, irrespective of the occurrence of injury or incidents. These alternative measures take various forms. For example, third-party audits have been used to measure the extent to which organizational safety management systems are compliant with preexisting standards. Other measurement approaches involve quantifying the direct causes of accidents, such as hardware failures or operational errors (Mohaghegh and Mosleh 2009), measuring the prevailing safety climate, and predicting safety behaviour and outcomes (Mearns et al. 2003; Neal and Griffin 2006).

Composite measures of workplace health and safety performance that combine traditional lag indicators with positive indicators of management activity and safety climate measures have been developed and used to evaluate the health and safety performance of large infrastructure construction projects (Lingard et al. 2011a, 2013a). Positive indicators of health and safety management activities have been labelled lead indicators. For example, Hopkins (2009a) states that 'lead indicators are those that directly measure aspects of the safety management system, such as the frequency or timeliness of audits' (p. 460).

Lead indicators of safety have also been described as 'precursors to harm that provide early warning signs of potential failure' (Shea et al. 2016). In the USA, Salas and Hallowell (2016) used lead indicators to develop a predictive model for providing early warning signs of changes in a construction contractor's safety management performance. These approaches show that lead indicators can be both positive (for example, management activity) or negative (for example, early warning signs). However, irrespective of whether they are positive or negative, the underlying logic is that measurement using lead indicators provides an opportunity to proactively manage workplace health and safety. Such measurement can guide responses to changes in the state of health and safety before incidents or injuries occur (Sinelnikov et al. 2015; Hinze et al. 2013).

6.6 What Leads and What Lags?

There is a great deal of inconsistency in the way the terms lead and lag are understood in relation to workplace health and safety metrics. Some consider the distinction between lead and lag indicators to lie in the position of the indicator in relation to the occurrence of harm, with lag indicators measuring harm directly and lead indicators measuring the precursors to harm. Others define lead indicators as practices that change before the actual level of risk people are exposed to changes, irrespective of whether harm eventuates (Kjellén 2009).

Alternatively, Hopkins (2009a) argues any kind of safety-relevant failure in a work system is a lag indicator. In this interpretation, the distinction between lead and lag indicators seems to depend on whether the indicator captures something positive (for example, the functioning of the safety management system) or negative (for example, the failure of a particular system defence or risk control mechanism). However, this interpretation presents challenges for models of safety incident causality in construction that identify causal factors in the immediate site environment, but trace these back to systemic factors in the project/organizational and external industry environments (see, for example, Haslam et al. 2003; Gibb et al. 2014). It is unclear whether organizational causal factors (for example, a poor design decision) should be regarded as a lag indicator, because they reflect a failure in a system's defences, or should be regarded as a lead indicator, because they constitute an early warning sign of a potential incident. How far back in the chain of causality does one need to go before a lag indicator should be viewed as a lead indicator?

Some argue the terms should be understood in a relative way, such that any event can be viewed as a lead or lag indicator depending on the perspective taken. For example,

Dyreborg (2009) suggests safety incidents may be considered as lag indicators of organizational safety performance, but as lead indicators when they are reported to a safety regulator and used to inform policies for prevention. Others criticise this relativist stance, arguing incidents that produce harm can never be regarded as lead indicators of safety performance (Hopkins 2009b). However, Hopkins (2009b) also criticises an absolutist approach in which the distinction between what is considered a lead or lag indicator is based on whether it occurs before or after an arbitrarily defined point in time. Take, for example, emergency procedures and systems designed to prevent or limit harm. Should they be classified as lag indicators because they come into effect after an incident has occurred, or as lead indicators because they are proactive measures of an organization's preparedness for safety incidents?

When used to describe indicators of workplace health and safety, the terms lead and lag are applied inconsistently. In the case example below (Case Example 6.2), we present an analysis of a five-year dataset that examines the relationship between some expected lead and lag safety indicators over time.

Case Example 6.2 Lead and Lag Indicators in a Five-Year Rail Infrastructure Construction Project

Data were collected as part of a routine reporting process implemented on a large infrastructure construction programme in Melbourne, Victoria. The multibillion dollar project took place over a five-year period and included construction of new rail track, new platforms at existing inner city train stations, new stations, a major upgrade to an existing suburban station, removal of two level crossings, 13 road and rail grade separations, and a new rail bridge.

Data were reported monthly to the client organization by the principal contractors using a standard event management system. Although this data was collected from multiple contractors undertaking different packages of construction work, the construction organizations supplying the data were contractually obliged to follow strict reporting requirements and collect standard safety performance metrics. The data were entered, verified, and collated by the client organization.

The Total Recordable Injury Frequency Rate (TRIFR) was selected as the presumed 'dependent' variable for the analysis. The TRIFR is a measure of the rate of recordable workplace injuries, normalized per million hours worked per year. While interpretation and reporting of Lost Time Injury Frequency Rates (LTIFRs) was likely to vary between contractors (see also Case Example 6.1), the reporting of total recordable incidents is less subject to differences in interpretation and manipulation.

The TRIFR was a statistically acceptable measure of safety outcomes due to the large number of person hours amassed at the project. The construction project involved a total of 14 593 250 worker hours. The number of worker hours per month was as high as 645 640, with an average of 239 234 worker hours per month.

The data were normalized (to control for variability in the number of employees and hours worked each month) and time stamped. The dataset was then analysed to explore temporal and causal relationships between expected lead indicators of safety performance and the TRIFR.

(Continued)

Case Example 6.2 (Continued)

Data was available for 61 reporting periods (months). Statistical techniques were used to detect significant relationships between safety performance indicators applying different time lags (for example, one month, two months, three months, and so on).

The results indicated some management activities measured at the project led changes in the TRIFR at subsequent points in time. For example, the frequency of 'toolbox' meetings led the TRIFR by four months, while pre-brief meetings and audits led the TRIFR by only two months. However, the statistical analysis also revealed that some management activities we expected to behave as lead indicators actually lagged changes in the TRIFR. That is, changes upwards or downwards in the TRIFR were significantly related to changes in the frequency of safety management activities at subsequent points in time. As an example, changes in the TRIFR were significantly correlated with subsequent changes in the frequency of alcohol and drug testing, the review of safe work method statements, site inductions, and safety observations.

This analysis suggests changes in the frequency of management actions can produce subsequent reductions in incident/injury frequency rates. However, the relationships are complex and reciprocal. An increase in incident/injury frequency rate also causes an increase in the frequency of safety management activity at a subsequent point in time. Therefore, the simple, one-directional relationship implied by the lag/lead terminology was not supported by the safety indicator data collected at this five-year project.

The analysis also provided evidence of cyclical relationships between safety performance indicators over time. For example, an increase in frequency of toolbox talks decreased the TRIFR in the short term. However, over a longer period, the direction of causality between these two indicators changed direction and a decrease in the TRIFR caused a subsequent decrease in the frequency of toolbox meetings.

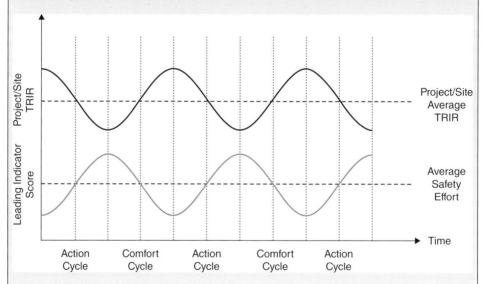

Figure 6.1 Theoretical simplified relationship between leading and lagging indicators. *Source:* Lingard et al. (2017a).

Case Example 6.2 (Continued)

These cyclical relationships may be indicative of a so-called incident cycle, in which it has been observed that managers respond to an increase in incident or injury rates by focusing greater attention on safety management practices. However, as the incident or injury frequency rate falls, so too can increased attention and emphasis placed on safety management in a workplace (Figure 6.1).

Stricoff (2000) observed a similar pattern, noting that: 'when the recordable rate exceeds a facility's upper-limit of perceived acceptability, management acts to drive the rate down. When the rate falls below that limit, attention to safety declines, and the recordable rate rises again. In this cycle, management action for improvement follows fluctuations in the injury frequency' (p. 37).

The 'knee jerk' reaction of management in the incident cycle described above also highlights practical problems inherent in relying too heavily on outcomes measures, such as the TRIFR, to measure workplace safety performance. This cyclical behaviour will not produce sustained improvement in safety performance over time.

(Adapted from Lingard et al. 2017a.)

6.7 Safety Climate Measurement

Many construction organizations have begun using safety climate surveys to understand the state of safety in the work environment. Safety climate is defined as 'a summary of molar perceptions that employees share about their work environments… a frame of reference for behaviours' (Zohar 1980, p. 96). Since Zohar's seminal paper, the concept of safety climate has been developed; Neal and Griffin (2006, pp. 946–947) define safety climate as 'individual perceptions of the policies, procedures and practices relating to safety in the workplace'. Safety climate represents workers' attitudes and perceptions of health and safety at a given point in time. It is distinguished from the organizational or project culture which refers to underlying core beliefs (Flin et al. 2000).

Cooper and Phillips (2004) suggest the concept of safety climate is important insofar as it predicts safety performance at a future point in time. Researchers have empirically investigated the relationship between safety climate and various aspects of safety-related behaviour and safety performance. Generally, but not always, the results have supported a link between safety climate and other aspects of performance. For example, on off-shore oil platforms Tharaldsen et al. (2008) report a significant inverse correlation between safety climate perceptions and incident rates. Varonen and Mattila (2000) similarly report that the incident rate in a sample of eight wood-processing companies was lower when the safety climate measures were high for dimensions such as organizational responsibility and safety supervision. These studies suggest safety climate may be considered a useful indicator of safety.

Some researchers have relied on self-reported measures of safety performance, again generally supporting a positive relationship between safety climate and performance. For example, Mearns et al. (2003) report that, in the offshore oil industry, favourable safety climate scores are associated with installations that have a lower proportion of self-reported involvement in safety incidents. Griffin and Neal (2000) and Neal and Griffin (2002) examined the relationship between safety climate and two types of self-reported safety behaviour: safety compliance and participation. They report that safety climate is positively related to both self-reported compliance with safety procedures, and to self-reported voluntary participation in safety-related activities, but that the strength of this relationship depends upon workers' levels of safety knowledge and motivation. Safety climate has also been linked to an organization's ability to appropriately attribute incident causes and learn lessons from safety incidents (Hofmann and Stetzer 1998). Evidence from longitudinal studies is also emerging to indicate safety climate measured at one point in time statistically predicts the occurrence of incidents or injuries at a subsequent point in time (see, for example, Wallace et al. 2006).

Consistent with research in other industries, there is empirical evidence to support a positive link between safety climate and the safety performance of construction organizations (Gillen et al. 2002). In Hong Kong, Siu et al. (2004) measured how construction workers perceived the safety responses of themselves, their colleagues, management, company safety officers, and their supervisors, reporting that aggregated safety climate scores were directly related to self-reported injury rate. Also in Hong Kong, Zhou et al. (2008) report that two climate dimensions (management commitment, and workmates' influence) exerted significantly greater influence on self-reported safety behaviour than workers' personal experiences of training and safety. In a lagged two-wave study of Swedish construction workers, Pousette et al. (2008) report that safety climate scores at one point in time significantly predicted self-reported safety behaviours seven months later (after controlling for prior levels of self-reported safety behaviour).

Despite the popularity and potential usefulness of measuring safety climate, there are a number of important considerations in applying, interpreting, and using safety climate data collected in construction organizations. These considerations relate to assumptions sometimes made about:

- the uniformity of the safety climate in a construction project or organization; and
- the stability of climate over time.

It is also important to understand the need to supplement safety climate data with information about the social context and operational environment of a construction organization or project in order to make sense of safety climate measurements and act upon them in an appropriate way.

In the remainder of Section 6.7 we will discuss these considerations and their implications for the design of safety climate assessment instruments and conduct of safety climate surveys.

6.7.1 Assumptions About Uniformity

Much of the safety climate research adopts the organization as the unit of analysis, implicitly assuming workers in construction organizations share a homogeneous perception of the priority placed on workers' health and safety by managers and others. However, there is growing recognition that workers develop perceptions of safety climate at different levels within organizations, and that the safety climate can vary significantly between organizational subunits (Zohar 2000).

Within a single organization there can often be significant variation in the quality of health and safety implementation between organizational subunits (Sparer and Dennerlein 2013). Thus, measuring the safety climate at the whole organization level can mask subtle but important differences that are relevant to local organizational health and safety performance.

Policies and processes at the organization level establish the context within which health and safety is enacted within organizational subunits (for example, in departments, projects, or workgroups). However, there is considerable scope for subunits in an organization to develop distinct characteristics. Zohar (2000) proposed two levels of safety climate:

1) that arising from the formal organization-wide policies and procedures established by top management;
2) that arising from the safety practices associated with implementing company policies and procedures within workgroups.

Zohar tested this proposition in a manufacturing context and confirmed that workgroup members:

(i) develop a shared set of perceptions of supervisory safety practices; and
(ii) discriminate between perceptions of the organization's safety climate and the workgroup safety climate.

Zohar suggests that group-level safety climates relate to patterns of supervisory safety practices or ways in which organization-level policies are implemented within each workgroup or subunit. Group-level safety climates are reported to influence workgroups' safety performance through shaping members' safety behaviour (Zohar 2002b). This means it is useful to measure the safety climate at different levels within organizations (Zohar 2008). Thus, individual climate scores are aggregated to the unit of analysis that is of interest. This can be the entire organization or organizational subunits, such as projects and workgroups (Zohar and Tenne-Gazit 2008).

In the highly fragmented construction industry context, differences between projects and (largely subcontracted) workgroups are likely to be even more significant. Construction projects are subsystems of an organization's larger portfolio of work. Each project is delivered through a temporary organizational structure in which professional services are brought in under a variety of contractual arrangements, and construction work is outsourced to a general contractor and a multiplicity of trade contractors. Uniformity of health and safety practices cannot be assumed within a single organization – work is highly decentralized and local managers (project managers and workgroup supervisors) necessarily exercise discretion in deciding how to implement organizational policies and procedures. Consequently, to understand the state of the safety climate in the 'projectized' construction industry, consideration should be given to characteristics of the organization, the project, and local workgroups.

Lingard et al. (2009b) tested whether Australian construction workers discriminated between group-level and organizational safety climates. They found that distinct workgroup safety climates were a feature of the Australian construction industry, and were driven by supervisors' and co-workers' actions and expectations about workplace health and safety. This means it is possible for the safety climate to vary at different levels within the same construction project. For example, workers may perceive:

• their (subcontracted) supervisors are strongly committed to health and safety (a group-level expression of climate); but

- senior managers in the principal contractor organization are less committed to health and safety (an organization-level expression of climate).

Some previously used safety climate measures include items relating to the organization (that is, top management and company policy) as well as subunit supervision. For example, in a safety climate survey of container terminal operators in Taiwan, Lu and Shang (2005) incorporate perceptions of supervisors' safety leadership. Fang et al. (2006) identified supervisors' and workmates' role as the third most important component of safety climate in the Hong Kong construction context. However, these researchers all aggregated these scores to the level of the entire organization. Similarly, the safety climate instrument developed by Jorgensen et al. (2007), and tested among a sample of English- and Spanish-speaking construction workers, combines questions about the general work environment (a useful indicator of the organization-level climate) with specific questions about workers' immediate supervisors (a group-level characteristic). We suggest the workgroup is a more appropriate unit of analysis for measuring supervisory and co-worker facets of safety climate.

Consistent with this view, Mearns (2009) argues that a single-level perspective inadequately reflects the state of health and safety within an organization because organizations are multi-level systems. Subcontracted workers are only loosely connected with the principal contractor and may work in a manner that is relatively isolated from their own company (Melia et al. 2008). This is likely to affect the development and impact of the safety climate, increasing the importance of measuring climate as a workgroup-level phenomenon. This is borne out by research conducted in high-risk industries. Both Findley et al. (2007) and Tharaldsen et al. (2008) report that, in nuclear decommissioning and in the offshore oil industries respectively, contracted workers have lower perceptions of safety climate compared to directly employed workers.

Lingard et al. (2010b) measured construction workers' perceptions of safety climate at various levels in an Australian building project. The results revealed that perceptions of the principal contractors' organizational safety climate were significantly related to workers' perceptions of subcontractors' organizational safety climate, as well as their workgroup supervisors' safety responses. But variations were still found between subcontracted workgroups, with some demonstrating a more positive and consistent orientation to workers' health and safety than others (Lingard et al. 2010c). Importantly, frontline managers and supervisors were identified as an important conduit through which senior managers' expectations about WHS were communicated to the workforce (Lingard et al. 2012b). This work was further developed by Zhang et al. (2015) who extended the multi-level safety climate assessment instrument used by Lingard et al. to include workers' perceptions of the safety commitment and leadership of client organizations in the construction industry. We discussed clients' ability to influence WHS in construction projects in Chapter 2.

6.7.2 Assumptions About Stability

With the exception of a few notable multiwave (longitudinal) studies, safety climate is mostly measured using a one-off cross-sectional approach. This assumes stability over time that is unrealistic in most work environments, but is particularly so in the constantly changing construction project context. Construction projects are characterized by constant change in the physical, social, and organizational environments. Different configurations of subcontractors are engaged at different times, and unexpected events and fluctuating workloads can produce pressure points that may change workers' perceptions of the relative priority placed on safety. There is emerging evidence to suggest the priorities people place on safety relative to other project goals can change over time (Humphrey et al. 2004).

In this environment, the one-off measurement of safety climate may not be useful because the state of safety measured at one point in a project may not reflect the state of safety across the project lifecycle.

Safety climate has been measured over the life of multiple construction projects delivered by a large food-manufacturing organization. This repeated, multi-level analysis of the safety climate provides benchmark data for safety climate at different points of construction completion, and helps to understand the dynamic changes in safety climate over the life of construction projects.

Longitudinal measurement of safety climate such as this can also be used to evaluate the impact of safety initiatives on construction projects. Figure 6.2 shows how, at project D, a targeted safety programme implemented in response to the climate survey at 47% construction completion was followed by a significant improvement in safety climate when measured at 68% construction completion.

6.7.3 Safety Climate Types

Zohar and Luria (2004) describe safety climate using two parameters: first, their strength, and second, their level.

Safety climate perceptions held by members of a particular social group (such as an organization, project, or workgroup) can range from weak to strong:

- In a strong safety climate there is very high consensus between members about the priority placed on safety.
- In a weak safety climate there is a low level of consensus concerning commitment to safety.

The level of the safety climate refers to the relative priority placed on safety within a group, as perceived by members of that group. The level of the safety climate can be expressed as either:

- high – that is, perceptions of a high level of safety commitment; or
- low – that is, perceptions of low safety commitment.

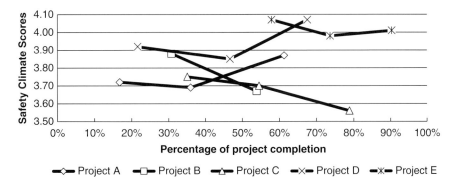

Figure 6.2 Safety climate in construction projects over time. *Source:* adapted from Zhang et al. 2018.

Table 6.1 Types of safety climate.

Type 1	An indifferent safety climate: • weak strength and low level
Type 2	An obstructive safety climate: • strong strength and low level
Type 3	A contradictory safety climate: • weak strength and high level
Type 4	A supportive safety climate: • strong strength and high level

It is possible for a safety climate that is supportive of safety (high in level) to be either weak or strong, depending upon the degree to which this perception is shared among workers in the same group.

Table 6.1 suggests four theoretically distinct types of safety climate positioned according to their strength and level.

Previous research in the Australian construction industry revealed that workgroups with supportive safety climates (i.e., those that are both strongly shared and high in level) had lower reportable and medical treatment injury rates than other workgroups (Lingard et al. 2010c).

6.7.4 The Need to Understand Safety Climate in the Social Context in Which It Occurs

Critics of safety climate measurement argue climate surveys can never reveal the deep and complex characteristics of organizational cultures which can only be understood through qualitative investigations. This may be the case, but quantitative measures of the safety climate are useful because workers are highly sensitive 'barometers'. They sense changes in managerial emphasis that can sometimes be subtle and not easily observed. Climate can, therefore, be a useful check on whether the behaviour of people in an organization matches the rhetoric.

However, any change in safety climate scores over time also needs to be explored to understand the underlying reasons for it. The state of the health and safety climate ascertained using questionnaire survey tools can provide important information about *what* is happening in an organization at a particular point in time. But understanding *why* health and safety are enacted in a particular way requires further probing. Therefore, it is recommended that the results of safety climate surveys are shared with workers and supervisors, and discussed in workshops, to 'unpack' their meaning and understand their significance. To be useful, the results of these surveys must be understood in the context of the social processes and meaning attributed to practices and events by people in the work environment (Antonsen, 2009b).

It is also important to pay attention to changes in safety climate scores and to understand the significance or drivers of them. Importantly, improvements over time should not necessarily be interpreted as representing major shifts in an organizational or project culture. This is because, unlike the deeper levels of organizational culture, safety climate is relatively malleable and easy to change (Naevestad 2009). Thus, if management commitment to workers' health and safety is being measured by a climate survey, it may be relatively easy to produce changes in safety climate scores by, for example, encouraging managers to discuss health and safety more frequently in their interactions with workers. It is much more difficult to change workers' underlying assumptions about health and safety.

6.8   Safety in Design Metrics

We have already discussed the argument for a move away from measuring the absence of safety to a more positive focus on measuring the presence of safety in a workplace or system of work. Most metrics currently used to measure health and safety in construction projects capture data during the construction stage of a project lifecycle. To date, very little emphasis has been placed on measuring the quality or effectiveness of health and safety management practices that occur before construction commences. It is widely acknowledged that decisions made during the pre-construction design and planning stages of a project are relevant to construction workers' health and safety. However, safety in design is not typically measured as a lead indicator of safety performance. Hale et al. (2007) echo this, arguing the current focus on analysing past incident reports to identify design as a causal factor fails to acknowledge the fact that, in many instances, design decisions are taken to proactively reduce WHS risks. Section 6.9 of this chapter proposes a new approach to measuring performance in planning and designing for construction workers' health and safety. This approach provides a positive performance indicator of the quality of pre-construction health and safety management.

The hierarchy of controls (HOC) is a well-established framework in work health and safety (WHS) (see, for example, Manuele 2006). The HOC classifies ways of dealing with WHS hazards/risks according to the level of effectiveness of the control. At the top of the HOC is eliminating a hazard/risk altogether. This is the most effective form of control because the physical removal of the hazard/risk from the work environment means workers are not exposed to it. The second level of control is substitution. This involves replacing something that produces a hazard with something less hazardous. At the third level in the HOC are engineering controls which isolate people from hazards. The top three levels of control (that is, elimination, substitution, and engineering) are technological because they act on changing the physical work environment. Beneath the technological controls, level four controls are administrative in nature, such as developing safe work procedures or implementing a job rotation scheme to limit exposure. At the bottom of the hierarchy, at level five, is personal protective equipment (PPE), which is the lowest form of control. Although much emphasized and visible on a worksite, at best PPE should be seen as a 'last resort'; see, for example, Lombardi et al.'s (2009) analysis of barriers to using eye protection. The bottom two levels in the HOC represent behavioural controls that seek to change the way people work. (For a summary of the limitations of these controls, see Hopkins 2006b.)

The HOC-based measurement approach provides a quantitative indicator of the effectiveness with which health and safety is being managed in the pre-construction stages of a project. It can be used to evaluate and compare design scenarios according to which would deliver the best WHS outcomes. Potentially, it could also be used by clients to specify or establish safety in design performance goals for a project.

We describe the HOC-based measurement approach in some detail below, providing a number of worked examples to illustrate its practical application.

6.9 Classifying Health and Safety Risk Controls Based on Their Effectiveness

The process recommended for using the HOC-based measurement method consists of five steps:

- *Step 1* – identify relevant 'features of work';
- *Step 2* – identify construction activities and tasks with health and safety implications;
- *Step 3* – categorize hazards associated with the construction activities;
- *Step 4* – identify risk control options for each of the hazards; and
- *Step 5* – classify and score the risk controls using the HOC.

These steps are explained below.

6.9.1   Step 1: Identifying Relevant 'Features of Work'

Construction projects can be divided into 'features of work'.[1] A feature of work is a group of activities which are distinct from other activities in terms of control requirements, location, work crews, or disciplines. Depending on the nature of the project, features of work could be based on the construction of specific structural elements (for example, constructing a cast-in-place concrete foundation or erecting steel columns), work breakdown structure (WBS) items, work packages (for example, pipe works, roof framing, or installing heating, ventilation, and air conditioning systems), or the project schedule (for example, erecting first-floor steel framing, second-floor overhead piping, and electrical services).

A feature of work should be defined narrowly enough to ensure adequate identification of WHS hazards and risk controls, yet not be so narrow that it overlooks hazards that may be not readily apparent.

6.9.2   Step 2: Identifying Construction Activities and Tasks with Work Health and Safety Implications

Each feature of work is broken down to identify the construction activities and tasks required for their construction and the significant WHS hazards inherent in these activities and tasks. This identification process should include people with appropriate construction experience and knowledge of construction processes and WHS.

6.9.3   Step 3: Categorizing Hazards Associated with the Construction Activities

Construction hazards are categorized according to their type (for example: falls from height slips trips and falls on the same level; struck by object or equipments; and coming into contact with a source of electricity).

---

1 The term is based on 'Defined Features Of Work' (DFOW) which is a terminology used by the US Army Corp of Engineers (USACE) and Naval Facilities Engineering Command (NAVFAC).

An appropriate WHS categorization scheme can be useful in this step, such as the National Institute for Occupational Safety and Health (NIOSH) Occupational Injury and Illness Classification System (OIICS) (Bureau of Labor Statistics 2012).

6.9.4   Step 4: Identifying Risk Control Options for Each of the Hazards

Identify ways to control the WHS risks posed by each hazard. This can either be ways an identified risk is to be controlled given the activities and tasks to be performed.

6.9.5   Step 5: Classifying and Scoring the Risk Controls Using the HOC

Score the selected or implemented risk controls according to the level of the HOC that they represent. Each control is given a score on a five-point scale, ranging from one (PPE) to five (elimination). In the event that no controls are identified, a value of zero is assigned.

Using this process generates an average HOC score for a particular feature of work. Thus, if two hazards are identified, one of which can be eliminated (score of 5) and the other controlled by administrative methods (score of 2), the average score would be 3.5.

The average HOC score reflects the quality and effectiveness of risk control solutions implemented for this feature of work.

Two worked examples are presented to illustrate the application of the HOC-based measurement method. These relate to measuring the quality of decisions made about the design of a high-rise building façade and about upgrading a sewerage treatment facility.

Worked Example: Assessing the Quality of Risk Controls for a High-Rise Building Façade System

The project used a design and construct delivery method in which the preliminary building design was completed by the client's architects and specialist consultants. The tender documents specified the building façade to be constructed of a lightweight frame structure made of glass reinforced concrete (GRC) with larger vertical sections made of precast reinforced concrete. During the tender process, the contractor raised concerns about the structural inadequacy of the GRC frame for a building of this height.

Following the engagement of the design and construct contractor, structural and constructability reviews were conducted to investigate design options and materials. A decision was made to use rolled steel sections instead of GRC elements. Consequently, façade members and connections were redesigned. Using much lighter steel elements reduced material handling and exposure to ergonomic hazards. It also eliminated the risk of the façade structure collapsing during or after construction.

The constructor proposed offsite manufacture of the façade. In this way, the construction process would be quicker. The need to store materials would also be eliminated and congestion on the small inner city site would be reduced. Offsite manufacture reduced exposure to the risk of contact with objects and equipment, and reduced the risk of falls, slips, and trips.

In the original planned sequence of work, the façade frame was to be fitted-off once the building structure was completed. However, the constructor suggested an

---

(Continued)

Worked Example: (Continued)

alternative sequence in which façade elements were to be fitted floor by floor as the building was being vertically constructed. This eliminated the need to work from swing stages or other mechanical equipment on the outside of the building. Workers could install and connect the framing beams from the finished floor levels in a safer manner.

The average HOC score for the revised design was 4.1. The total HOC score was 61 across 15 identified safety challenges. The majority of these challenges were managed using high level technological control measures producing this high average HOC score. Table 6.2 presents the definition of activities, tasks, safety challenges and responses, and HOC-level values used to generate the overall average score.

Table 6.2 Assessing the quality of risk controls for constructing a high-rise building façade system.

Activity	Work task	Safety challenge	Response to safety challenge	HOC level	HOC score
Material handling and construction activities for the WRAP façade	Installation of horizontal frame elements for the façade structure	Overexertion in holding, carrying, or wielding Struck, caught, or crushed in collapsing structure, equipment, or material	Use lightweight material to build frame elements	Substitution	4
Installation of frame elements for the WRAP structure (façade)	Connecting the frame elements back to the slab	Overexertion bending, crawling, reaching, twisting, climbing, stepping	Use rolled steel in place of GRC and reduce the number of connections required	Substitution	4
Building WRAP frame elements	Building façade frame elements from rolled steel folded into rectangular shape	Contact with objects and equipment Overexertion in holding, carrying, or wielding	Offsite manufacturing	Elimination	5
Installation of steel elements	Lifting large sections to position using crane	Struck by object or equipment	Training, safe work method statement, work sequence	Administrative	2
Installation of façade frame	Positioning and connecting frame elements to each other and to the slab	Falls to lower level	Install façade elements floor by floor, accessing the work area from finished floors	Elimination	5

Worked Example: (Continued)

Table 6.2 (Continued)

Activity	Work task	Safety challenge	Response to safety challenge	HOC level	HOC score
Installation of façade frame	Installation of façade frame elements at each floor without permanent exterior walls	Falls to lower level	Protection by safety screens	Engineering control	3
Installation of façade frame elements	Connecting the intersecting elements together	Overexertion bending, crawling, reaching, twisting, climbing, stepping	Fabricate the intersecting sections offsite as a single section to reduce the number of connections	Substitution	4
Fixing façade frame to the slab	Connecting the frame back to the slab to fix the façade	Contact with objects and equipment	Cast ferrules into the precast slab to eliminate the need for drilling into the concrete	Elimination	5
Beam connections	Connecting the beams to the intersecting sections using connection arms	Overexertion bending, crawling, reaching, twisting, climbing, stepping	Attach connection arms to the beams in factory to eliminate the need to weld or bolt the connection arms onsite	Elimination	5
Frame connections	Connectors between frame and cast-in ferrules	Overexertion bending, crawling, reaching, twisting, climbing, stepping	Use connectors providing 20 mm tolerance in all directions to provide some flexibility during installation	Substitution	4
Beam connections	Installing and tightening bolts on connection plates inside the beams	Overexertion bending, crawling, reaching, twisting, climbing, stepping	Increase the size of the panel openings to have more space and better access to the connection area	Substitution	4

(Continued)

Worked Example: (Continued)

Table 6.2 (Continued)

Activity	Work task	Safety challenge	Response to safety challenge	HOC level	HOC score
Beam connections	Installing and tightening bolts on connection plates inside the beams	Falls to lower level	Access to all connection points specifically located in a position easily reached from finished concrete floors. Clearance between the façade frame and the building reduced to allow frame connection works to be undertaken from behind the safety of perimeter barricading.	Elimination	5
Vertical frame elements	Temporary works to install precast reinforced concrete vertical elements spanning two floors	Struck, caught, or crushed in collapsing structure, equipment, or material	Prop the vertical elements into position to resist wind and lateral forces while waiting for the next floor slab to be ready to continue installation	Engineering control	3
Vertical frame elements and connections	Connection between vertical elements and crisscross sections on top levels	Contact with objects and equipment	Design the vertical precast elements to span two floors to reduce both the number of connections required and the amount of temporary works needed to support the elements	Substitution	4
Painting the frame	Painting the frame	Falls to lower level	Paint the elements prior to installation; only touch-ups were done onsite in case of any damage	Substitution	4

Worked Example: Assessing the Quality of Risk Controls for Upgrading a Sewerage Treatment Facility

An existing centrifuge and existing piping were to be upgraded at a sewerage treatment plant. The new equipment was to be connected to existing live piping infrastructure; however, to install the equipment a number of existing pipes would need to be removed. As the majority of the pipes were suspended from the ceiling, this work was to be carried out at height using elevated working platforms or scaffolding.

During the design stage it was found that the new centrifuge would need to be placed over a large void cut into a suspended slab. The void provided a connection to the inflow and outflow piping system. The existing centrifuge was larger than its replacement. Thus, to install the new centrifuge activities such as infilling part of the opening to make it smaller or constructing some type of supporting system to span the void would be necessary, introducing new hazards to the construction process.

During procurement it was also discovered that the new centrifuge would not meet capacity requirements stipulated by the client/operator. Consequently, a larger centrifuge that met capacity requirements and that was safer to install was purchased. This centrifuge was to be located on a mezzanine level with an adjoining void equal to the height of a six-storey building. During installation of the centrifuge it was identified that, due to its size, full perimeter access around it was not possible and that a platform would need to be installed. This involved connecting a steel platform to the edge of the concrete mezzanine floor and cantilevering over the void. Installing the platform would prevent workers from having to lean out over the void to gain access to the end of the centrifuge. While a large portion of the platform was erected offsite, access to the edge of the slab was still needed to fix the platform into position. A specialist scaffolding contractor was engaged to design and install a temporary cantilever scaffold to address hazards associated with working from this height. Due to the size and weight of the partially completed platform, a crane was used to move the structure into position; however, existing plant and infrastructure in the area severely hampered the crane's movements. Other WHS hazards were also identified with this work, including effects of fumes and gases in carrying out onsite welding.

One control strategy used to address these risks was wearing PPE. Given that the work was carried out during the summer months and within close proximity to an industrial heater, the use of PPE to mitigate the identified risks produced new hazards, such as heat stress and fatigue.

The average HOC score for the design was 2.9. Table 6.3 presents the definition of activities, tasks, safety challenges, and responses, and HOC-level values used to generate the overall average score. The total HOC score was 29 across ten identified safety challenges. Proportionally more of these challenges were resolved using behavioural measures and fewer challenges could be resolved using high level elimination or substitution strategies. This contributed to the relatively low HOC score for this design.

(Continued)

Worked Example: (Continued)

Table 6.3 Assessing the quality of risk controls for upgrading a sewerage treatment facility.

Feature of work	Activity	Work task	Safety challenge	Response to safety challenge	HOC level	HOC score
Installation of centrifuge	Fitting and installation of the centrifuge over the slab opening	Temporarily suspending the centrifuge over the opening using a crane to modify the supports and fittings, due to difference in size of the new and the old centrifuges	Struck by object or equipment Struck, caught, or crushed in collapsing structure, equipment, or material	Change the centrifuge type to fit over the slab opening	Elimination	5
Pipe works	Installation of temporary pipes to connect the centrifuge to the existing infrastructure	Working around existing pipes and structures, carry, lift, and connect pipes. Remove and reinstall existing pipes in some cases	Struck by object or equipment	Use safety hats and gloves	PPE	1
Pipe works	Upgrading the existing piping system	Access to pipes suspended from ceiling	Falls to lower level	Elevated platforms and scaffolding	Engineering control	3
Pipe works	Connections	Welded connections	Ignition of clothing from controlled heat source Exposure to harmful substances or environments	Use 'Vitolux', no need for welding, and easy and quick to install	Substitution	4

Task	Activity	Description	Hazard	Control	Control type	Risk
Installation of centrifuge	Access around the centrifuge	Workers lean out over the adjoining void to gain access to end of the centrifuge	Falls to lower level Overexertion bending, crawling, reaching, twisting, climbing, stepping	Install a steel platform to the edge of the concrete slab cantilevering over the void	Engineering control	3
Construct/erect the steel platform	Steel works	Steel works to erect the platform, onsite vs. offsite	Contact with objects and equipment	Offsite manufacturing	Elimination	5
Install the platform	Installation works at height	Installation works at height	Falls to lower level	Temporary cantilever scaffolding	Engineering control	3
Install the platform	Lifting	Lifting the prefabricated platform into position using a crane, close to existing infrastructure	Struck, caught, or crushed in collapsing structure, equipment, or material	Safe work method statements, job training, work sequencing	Administrative	2
Install the platform	Welding	Onsite welding to install platform	Ignition of clothing from controlled heat source Exposure to harmful substances or environments	Use protective equipment	PPE	1
Install the platform	Working in summer close to an industrial heater	Working in summer close to an industrial heater and wearing PPE	Exposure to temperature extremes Overexertion and bodily reaction	Induction, job rotation	Administrative	2

6.10 Using HOC Method for Comparison

The following example illustrates the use of the HOC-based measurement method for evaluating the effectiveness of proposed design changes to control WHS risks during excavation work.

Case Example 6.3 Assessing the Quality of Health and Safety Risk Controls for Excavation Activities in Constructing a Basement Mausoleum

A basement mausoleum was to be constructed in a cemetery. The site was surrounded by existing graves with established trees planted among them. To maximize the usable area, the client proposed a setback of just over 2 m from the adjoining grave sites and trees.

The temporary works design required that a retaining wall and bored concrete piles be constructed, at 1800 mm centres, around the perimeter of the excavation to retain the soil. External propping using ground anchors was then to be installed to prevent rotation of the wall. The exposed soil between the piles would then be retained using shotcrete. Once the temporary works were completed construction of the permanent works could commence from the bottom up.

However, once engaged, the constructor proposed a safer top-down approach in which construction of a retaining system would start at ground level and progressively work its way down as excavation continued in stages, until the required depth was reached. The constructor also proposed eliminating the rock anchors due to a number of risks associated with them. To ensure the anchors posed no threat to any construction activities that may occur next to the mausoleum in future, the ground anchors would need to be de-stressed. In the original design, gaining access to the anchors to de-stress would require the constructor to enter the 'gap' between the temporary wall and the mausoleum wall, remove the anchor's cap and then destress or cut the steel rods in a small, confined space. This would create ergonomic hazards for workers having to manoeuvre within a confined space. The potential for the stressed bars to react and hit the workers when released created additional WHS risk.

The internal propping required for the system had to be designed to provide enough clearance for the machinery to move safely around without the danger of running into and knocking over props. To achieve this, the constructor proposed to use 'Megaprops' which are unlike alternative internal propping systems that connect to the face of the wall and are anchored back down into the bottom of the excavation, taking up a lot of valuable space. Megaprops are large steel beams installed at the top of the excavation which span the width of the excavation, pushing back against opposing walls. This requires fewer props to be installed and frees up the base of the excavation so that a clear and unobstructed area is available to undertake excavation.

For ease of installation, the connection brackets were cast on to the top of the ring beam rather than on the walls. This eliminated the need to drill into the concrete at a later stage to secure the props. To assist with the Megaprops installation, each connection plate was made with a 'lip' that provided temporary support to the props once they were lowered onto the connection plate. The connection bolts could then easily be threaded through the prop and into the connection plate without the need for a crane to hold it in

Case Example 6.3 (Continued)

position, until such time as the prop was fixed at both ends. All fixing could be done at ground level due to the connections being located on the top of the capping beam.

Table 6.4 shows the application of the HOC evaluation method to the mausoleum case study. The average HOC score is calculated. (Only tasks related to excavation of the basement are included.)

Table 6.4 compares the effectiveness of WHS risk controls before and after the changes proposed by the constructor.

The average HOC score for the original design was 2.1. The average HOC score for the revised design was 4.3. This shows that the revised design produced more effective (technological) controls for identified WHS risks.

The differences between the risk control profile of the original design and the revised design are visually presented in Figures 6.3 and 6.4. Graphical presentation of the distribution of HOC scores associated with a particular design solution can be particularly useful for design professionals to quickly assess the spread of effective health and safety risk control solutions. By mapping the implemented risk control solutions before and after design changes, design professionals can evaluate the effectiveness of these changes for controlling workplace health and safety risks (Lingard et al. 2015c).

Table 6.4 Evaluation of health and safety risk controls for the basement excavation.

Task	Hazard	Original design solution	Original HOC level and score	Average HOC score	Revised design/ OSH intervention	Revised OSH control level and score	Average HOC score
Excavation using small machinery	Struck by object or equipment	Establish exclusion zones, appointing spotters	Administrative (2)	2.1	—	—	4.3
Deep excavation (8.5 m)	Caught in or compressed by equipment or objects	Temporary works to retain the soil	Engineering control (3)		—	—	
Install temporary works in the excavation ditch	Caught in or compressed by equipment or objects	Bored concrete piles, propping, shotcrete (trained workers working in the excavation ditch)	Administrative (2)		Top-down excavation and installing temporary works simultaneously. No temporary work after excavation	Elimination (5)	
Temporary works. Propping inside the excavation ditch	Struck, caught, or crushed in collapsing structure, equipment, or material	Trained workers enter the excavation ditch and install props	Administrative (2)		Install Mega- props. No need to enter the ditch. Workers to install Mega- props from ground level	Substitution (4)	

(Continued)

Case Example 6.3 (Continued)

Table 6.4 (Continued)

Task	Hazard	Original design solution	Original HOC level and score	Average HOC score	Revised design/ OSH intervention	Revised OSH control level and score	Average HOC score
Excavation using machinery	Caught or compressed by collapsing material	Machinery working close to props, appointing spotters to avoid hitting props	Administrative (2)		Use Megaprops. No need for props in the excavation ditch	Elimination (5)	
Destressing the rock anchors	Struck by object or equipment	Trained workers remove the anchor's cap and then destress or cut the steel rods	Administrative (2)		Use Megaprops. No need for rock anchors	Elimination (5)	
Destressing the rock anchors	Working in a confined space	Trained workers enter the 'gap' between the temporary wall and the mausoleum wall	Administrative (2)		Use Megaprops. No need for rock anchors	Elimination (5)	
Temporary works, installing Megaprops	Fall from height Overexertion in holding, carrying, or wielding	Form work around the brackets as well as sealing	Administrative (2)		Cast brackets on to top of capping beam. No need for installation	Elimination (5)	

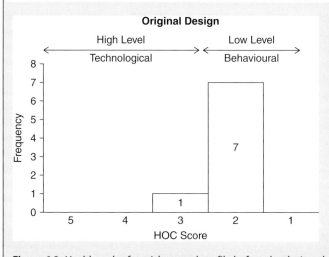

Figure 6.3 Health and safety risk control profile before the design change.

Case Example 6.3 (Continued)

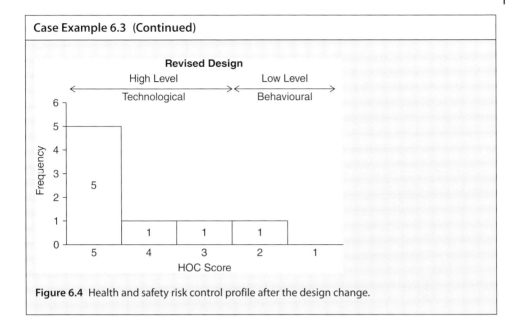

Figure 6.4 Health and safety risk control profile after the design change.

The HOC-based method can also be used to help to decide on, change, and improve health and safety risk control solutions at early stages of projects, monitor and review them during design development, and communicate reasons for design decisions to other project stakeholders. As the graphs show, the original design relied on workers' behaviour and onsite controls (lower-level controls) to address health and safety risk, with the majority of risk controls being behavioural/lower-level controls.

Using this HOC-based measurement method can help compare different design options in terms of the quality and effectiveness of their WHS outcomes. Further, the method identifies features of work with lower-level controls so interventions to improve these risk controls can be implemented.

The HOC provides a framework for eliminating or controlling hazards. It enables decision makers to consider the effectiveness of different control measures for WHS risks, helping them to achieve the best level of workplace health and safety. The HOC-based method offers a numerical system that provides a way of quantifying and comparing control options for various features of work.

Using this method should assist stakeholders in construction projects to gain a better understanding of WHS hazards and related control measures. Over time, using a HOC-based indicator to measure, evaluate, and benchmark safety in design decisions will encourage people to focus on health and safety solutions that have longer-lasting benefits for improved constructability and workers' health and safety. That orientation is far preferable to providing quick fixes that do not capitalize on opportunities to improve workers' health and safety by eliminating risks from, or engineering risks out of, work processes before construction work commences.

6.11 Conclusions

In this chapter we considered different types of indicator currently being used to measure workplace health and safety performance. The review shows there is no single perfect indicator – and all indicators in current use have limitations and need to be interpreted with these in mind. Some people call for less measurement because the wrong things are being measured, which drives behaviours that do not produce safer, healthier workplaces. Certainly, our analysis of both lag and so called lead indicator data does show managers can become overly focused on managing the metrics – particularly when these metrics underpin the health and safety aspects of commercial frameworks used to deliver construction projects.

There is presently a focus on using a discrete set of metrics to measure health and safety as a stand-alone facet of project or organizational performance. This is unhelpful because the resulting analysis does not provide an understanding of the emergence of health and safety in the broader organizational, technological, and social contexts of construction projects.

To properly understand an organization's health and safety performance, a broader set of indicators is required because health and safety does not occur in isolation. Other factors in the broader project environment are also important, and are potentially more meaningful indicators of health and safety performance than frequency counts of specific health and safety management activities. Thus, linking health and safety data to other project performance data is suggested, as is design and development of appropriate indicators of the extent to which health and safety is integrated into 'upstream' (that is, planning and design) decision making in construction projects.

Opportunities exist to measure the level of health and safety risk in a workplace more directly, rather than rely on 'after the fact' measures or indirect measures of management activity. Advances in sensing technology, machine learning, and big data analytics provide opportunities to collect large volumes of data about project events, the physical work environment, and workers' health and safety perceptions and experiences. Previous research has tried to develop predictive models of safety failure. However, the critical opportunity now lies in understanding project information related to health and safety success. Combined with real-time data collection, the development of predictive models would establish preconditioning factors for health and safety success in construction projects.

It is also important that any further development of measurement tools and predictive models place greater emphasis on measuring occupational and environmental health risk factors and outcomes, and on identifying precursors to good health in construction workers.

Discussion and Review Questions

1 How useful are injury or incident rates as measures of organizational and/or project-level WHS performance?

2 What alternative measures of workplace health and safety are there?

3 How useful are the terms 'lag' and 'lead' in relation to measuring workplace health and safety?

4 Should the construction industry change the way health and safety performance is measured and evaluated? If so, how?

Acknowledgements

Some of the research presented in this chapter was supported by Cooperative Agreement Number U60 OH009761, under which RMIT was a subcontractor to Virginia Tech, from the USA's Centers for Disease Control and Prevention (CDC), National Institute for Occupational Safety and Health (NIOSH). The authors also gratefully acknowledge funding provided by the Major Transport Infrastructure Program, Department of Economic Development, Jobs, Transport and Resources, Victorian Government.

7

Managing Work Health and Safety: Rethinking Rules and Worker Engagement

Helen Lingard, and James Harley

School of Property, Construction and Project Management, RMIT University, Melbourne, Victoria, Australia

7.1 Introduction

7.1.1 Managing Work Health and Safety

The latter part of the twentieth century saw increased focus on systematic work health and safety (WHS) management. Some countries mandated implementing a systematic approach to WHS management; see, for example, the notion of internal control in Norway and Sweden (Gaupset 2000). Other countries have adopted a hybrid regulatory regime. For example, Saksvik and Quinlan (2003) describe how Australian organizations have 'voluntarily' adopted occupational health and safety management systems (OHSMSs) in response to a shift from prescriptive to process-based forms of regulation. Importantly, Frick and Wren (2000) draw a distinction between legislated systematic WHS management and (usually voluntary) implementation of formal and documented WHS management systems. This chapter will reveal this distinction as important because formal WHS management systems have been subject to considerable review and criticism.

Gallagher et al. (2003) define a WHS management system as '... a combination of the planning and review, the management organisational arrangements, the consultative arrangements, and the specific programme elements that work together in an integrated way to improve health and safety performance' (p. 69). Attempts have been made to identify the elements of an effective WHS management system (see, for example, Redinger and Levine 1998). In 1997, Australia and New Zealand were among the first countries to develop a guidance standard on WHS management systems, followed by a certification standard in 2000. Standardization has continued and an international standard (ISO 45001:2018 *Occupational health and safety management systems – Requirements with guidance for use*) was recently released.

Market-based activity has also driven implementation of WHS management systems as consultants have sold proprietary systems to corporations (Frick and Wren 2000). There are often institutional and/or commercial pressures to implement WHS management systems. For example, since 2004, Australian legislation has linked WHS

Integrating Work Health and Safety into Construction Project Management, First Edition.
Helen Lingard and Ron Wakefield.
© 2019 John Wiley & Sons Ltd. Published 2019 by John Wiley & Sons Ltd.

performance and practices to tendering opportunities for publicly funded construction projects. Under the Australian Government Building and Construction Occupational Health and Safety Accreditation Scheme ('the Scheme'), head contractors awarded construction work funded directly or indirectly by the Australian Government (above a threshold value) must be accredited. Accreditation involves submitting an application, followed by a process of onsite auditing of a construction company's WHS management practices and performance.

Concerns have been raised about the growing proceduralization of WHS, and the reliance of management systems on rules and paperwork. It is argued that critical information needed by workers is often 'buried' inside long and overly complicated documents (Bieder and Bourrier 2013). Despite their length and complexity, formal WHS documents record 'work as imagined' by managers rather than 'work as performed' (Borys 2012b). Paperwork related to WHS management has been criticized for reducing both managers' proximity to work and opportunity for 'hands on' WHS management (Lamvik et al. 2009).

Organizations that buy WHS management system packages 'off the shelf' are often unaware of the embedded assumptions inherent in these systems and/or the need to adapt them to specific organizational conditions (Saksvik and Quinlan 2003). Reiman and Rollenhagen (2011) argue that WHS management is always based on underlying theories about people, organizations, and WHS. Depending on the perspective taken, certain issues will be emphasized and particular solutions will be preferred.

The flavour and emphasis of a WHS management system has been linked to broader theories of organization and management. Nielsen (2000) argues that WHS management has been strongly influenced by rational theories about management and organizations that stress structures, goal attainment, and efficiency. The focus on formalization, prescription, and measurement can be traced back to Taylor's notions of scientific management and classical management theories that emphasize top-down control of work processes and practices (Nielsen 2000). In the same vein, Reiman and Rollenhagen (2011) suggest WHS management systems often reflect traditional mechanistic approaches which favour structures and emphasize the control of behaviour through establishing prescriptive procedures and instructions, supported by enforcement and supervision.

Some audit tools used for WHS management systems have also been identified as potentially counterproductive, reducing WHS to a paper-based, 'tick and flick' practice. Hohnen and Hasle (2011) argue that reducing WHS to observable (and therefore auditable) elements ignores the softer aspects of organizational cultures, management commitment, and worker consultation, each of which are important for WHS performance (see also Chapter 5 on developing an enabling culture for WHS). Worse still, Hohnen and Hasle identify the risk inherent in the need to demonstrate compliance in WHS management systems audits – a compliance emphasis focuses attention on the audit process as an end in itself, potentially detracting from more important questions of how well risks are being controlled, or how WHS outcomes can be further improved.

7.1.2 Contrasting Viewpoints About How to Achieve WHS

In his book *Ten Questions about Human Error* (Dekker 2005), Sidney Dekker identified two contrasting approaches to managing safety, which he refers to as model 1 and model 2. A model 1 approach regards workers as having limited competence and expertise, as compared to managers, professionals, and technical experts. Because workers are fallible, their behaviour must be controlled through establishing rules that prescribe how work is to be performed. These action rules, based on an analysis of task and risks, are written by experts with (assumed) higher levels of competence and expertise than workers. These formal rules are applied inflexibly and intended to comprehensively specify how work is to be done in all possible work situations and scenarios. A model 1 approach is based on principles of scientific management and reflects a rationalist epistemological position which holds that a single best way of working can be identified and prescribed. Model 1 approaches rely heavily on documenting WHS in formalized procedures, which are updated only infrequently when work processes change. When accidents occur, a model 1 response will involve identifying behavioural causes of the accident and creating additional rules to prevent a recurrence. As we discuss below, a model 1 approach focuses on enforcement, and any failure to comply with rules is framed as ignorant or deviant behaviour that should be understood in order to stop it.

In contrast, a model 2 approach recognizes workers' experience and competence in performing their work. This approach acknowledges rules cannot cover all eventualities. Neither should they be too rigidly applied because they are, by necessity, adapted to suit localized and situated ways of working. This approach recognizes that inevitable variability introduced by humans engaged in work practices can have positive safety impacts, as people adapt their practices to suit their environmental conditions. This variability is only regarded as a problem if it approaches too closely a boundary of acceptable or tolerable practice. A model 2 approach sees workers (rather than managers, professionals, and technical specialists) as being experts, and rules are reframed as resources that support, rather than constrain, action. In a model 2 approach, rule breaking is not always regarded as undesirable, and when rules are broken the violation is often traced back to a gap between the way a rule is framed and the reality of the work situation.

Although these descriptions are somewhat exaggerated and caricatured, they nonetheless reflect assumptions underpinning the way organizations (and their managers) approach WHS. In some instances, organizational WHS policies and practices can reflect contradictory logics. For example, research by Sherratt et al. (2013) contrasts the content of organizational policy documents and public statements about worker engagement with site-based practices that seek to control workers' behaviour through establishing and enforcing formal rules.

In this chapter, we explore different perspectives relating to human failure, and consider the implications of these perspectives for WHS management, particularly as it relates to establishing and managing rules. We examine the causes of rule violations and question the view that rules are broken by deviant workers. Instead, we discuss organizational and environmental factors involved in example cases of rule breaking in the construction industry. We discuss the importance of rule management as an approach to managing WHS that establishes important action rules, yet also supports using workers' knowledge and experience to develop safer and healthier ways of working. Finally, we describe a case study that used a novel participatory video approach to engaging workers in designing work procedures that better reflect situated work practices. We describe how workers' ideas for WHS improvements were 'unlocked' and included in the development of visual WHS procedures that are being used for educational and instructional purposes in the Australian construction industry.

7.2 What Is Human Error?

Human error is frequently identified as an important factor in workplace safety incidents/accidents in construction and other industries. See Garrett and Teizer (2009) for a discussion of the classification of human error in construction safety incidents/accidents. However, these analyses often use the term 'human error' uncritically. Hollnagel and Amalberti (2001) reflect that the popularity and widespread reference made to the term 'human error' can be attributed to its apparent simplicity. Yet, they argue, there is no clear definition of human error and the term, in fact, means different things to different people. Thus, human error is sometimes used to describe the cause of something, an event, or the outcome of an action (Hollnagel and Amalberti 2001). Dekker (2002) similarly argues that error can be seen as the cause of a failure (that is, an event is due to human error), as the failure itself (for example, a particular behaviour or action was in error), or a process (that is, the error is a departure from some kind of standard). In the latter framing, which actions are determined to be errors will depend upon the standard that is applied to a particular situation.

According to guidance developed by the UK's Health and Safety Executive, a human error is 'an action or decision which was not intended, which involved a deviation from an accepted standard, and which led to an undesirable outcome' (Health and Safety Executive 1999, p. 13). Thus, this definition reflects elements of the process and consequence, but also includes an element of intention; that is, an error, by this definition also involves unintended behaviour.

The problems associated with the use of the term human error were observed by Rasmussen (1982), who commented that:

> Frequently they (human errors) are identified after the fact: If a system performs less satisfactorily than it normally does – due to a human act or to a disturbance which could have been counteracted by a reasonable human act – the cause will very likely be identified as a human error.
>
> *(p. 313)*

This reflects the challenge that attributing human error is always a 'judgement in hindsight' (Hollnagel and Amalberti 2001). That is to say, an action may only be considered to be in error if it is deemed to be so after the event, usually because an undesirable consequence has prompted an investigation. The problem of backward causation refers to trying to identify causes of events after they have happened, which presents significant challenges for reasoning and logic. Because two events (an action and an outcome) occur contiguously, it does not logically mean they are causally related. The premise that 'What You Look for is What You Find' has been observed in incident/accident investigation; that is, if human errors are sought they will likely be found (see, Lundberg et al. 2009), particularly when human factor error classification systems for classifying causes are used to understand the human contribution to incidents/accidents (Dekker 2002).

Dekker (2002) is particularly critical of 'after the fact' methods for classifying human errors. According to Dekker, these are:

- highly subject to hindsight bias;
- based on judgement rather than analysis; and
- do little to explain why people acted as they did, given the circumstances in which they found themselves (Dekker 2002).

In relation to the latter point, the circumstances surrounding the error are almost always much more complex than error classification systems would suggest and can include, for example:

- competing organizational priorities and conflicting goals;
- resource and time constraints;
- limitations associated with equipment or technologies;
- information overload;
- communication breakdowns; and/or
- interpersonal or coordination failures among team members.

Dekker (2002) argues that:

> … the point in learning about human error is not to find out where people went wrong. It is to find out why their assessments and actions made sense to them at the time, given how their situation looked from the inside.
>
> *(p. 8)*

Hollnagel and Amalberti (2001) also suggest it is overly simplistic to consider behaviour as being either right or wrong, correct or incorrect, because people may subconsciously or consciously adjust or compensate for their actions before any consequence occurs. Where actions perceived as not being carried out correctly are detected and corrected, actual and intended outcomes are the same and the actions should be considered to be correct. If, on the other hand, a system is unforgiving to the extent that actions perceived to be incorrect are detected but recovery is not possible, actual and intended consequences do not match and the action may be considered an error. In yet another scenario, actions that are perceived to be incorrect, and are detected, may be ignored. Hollnagel and Amalberti (2001) suggest this occurs most commonly when a person assesses the expected consequences of the action to be unimportant. In this case, whether the action is regarded as an error or not is likely to be determined by the seriousness of any consequences that flow from the action.

7.3 Human Error Types

Notwithstanding problems inherent in applying the term 'human error', it remains widely used. Various classification systems have drawn a distinction between different types of human error. Perhaps the most well-known of these was developed by Rasmussen (1982), who proposed a generic psychological classification for human errors that specified relations to particular task properties and environmental characteristics. Rasmussen distinguished between three levels of behaviour: skill-based, rule-based, and knowledge-based behaviour:

- Performance of skill-based behaviour, which includes automated, subconsciously executed routines, is controlled by stored patterns of behaviour in a time–space domain. Errors occur as a result of variation in force, time, or space coordination.
- Performance of rule-based behaviour occurs in familiar situations controlled by stored rules for coordinating subroutines. Errors can occur when people wrongly classify or recognize situations, make erroneous associations relating to the task, or fail to recall procedures. Rasmussen (1982) notes that rule-based behaviour is used to control skill-based routines and, thus, error mechanisms related to skill-based behaviour are always active. According to Rasmussen (1982), rule-based behaviour is goal oriented, rather than goal controlled, meaning criteria for error depend on whether relevant rules are recalled correctly and applied.
- Performance of knowledge-based behaviour occurs in unique or unfamiliar situations in which actions need to be planned based on knowledge of the functional, physical properties of a system and priority of goals.

Rasmussen explained how these three types of behaviour are driven by different information processes, each presenting distinct mechanisms for human error. James Reason (Reason 1990) also categorized errors in terms of whether they are skill-based slips and lapses, rule-based mistakes, or knowledge-based mistakes. This classification system has been adopted in guidance on human factors and error reduction (see, Health and Safety Executive 1999) and is illustrated in Figure 7.1. Violations are distinct from error and their causes will be discussed later in the chapter (see Section 7.7).

According to the classification system, skill-based errors can occur when people are distracted or preoccupied with things other than the task, leading to slips or lapses. Slips and lapses generally occur when people are performing very familiar tasks (for example, driving a car), which are carried out without much need for conscious attention. Even very skilled and experienced workers are prone to slips and lapses if their attention is diverted from the task they are performing.

Slips are 'actions-not-as-planned'; for example, omitting a step in a work sequence. But lapses occur when someone forgets to carry out an action, loses their place when performing a task, or perhaps forgets what they intended to do. Lapses happen when people are distracted.

Rule-based and knowledge-based errors are referred to as 'mistakes'. These are deliberate actions taken by people who do the wrong thing believing it to be right (Health and Safety Executive 1999). Mistakes differ from slips and lapses in that they are not necessarily related to inattention or distraction, but reflect a failure in mental processes. A rule-based mistake can occur, for example, when a set of rules is remembered but wrongly applied to a situation. A knowledge-based mistake occurs when a problem or situation is unfamiliar, misdiagnosed, and the wrong action is applied.

Table 7.1 provides examples of the different types of error related to using mobile elevated work platforms in the construction industry. It is evident from these examples that strategies to prevent human failure need to focus on the type of failure they are designed to prevent. Slips and lapses associated with using mobile elevated work platforms have been reported when the joystick control on one model of a scissor lift is jointly used for both the lift and drive functions. To change functions the operator

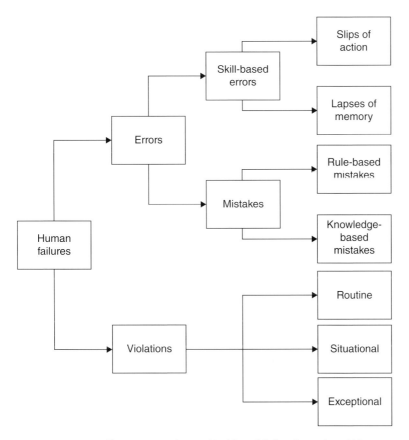

Figure 7.1 Types of human error. *Source:* Health and Safety Executive 1999.

is required to change the button below the joystick control to select either the lift or drive function. Leah et al. (2013) report instances of experienced operators either:

- moving the joystick in the wrong direction from that intended, with the drive/lift function selected correctly (a slip); or
- forgetting to carry out the action of changing between the lift and drive function before operating the joystick (a lapse).

Thus, slips and lapses could potentially be reduced by modifying the design of the controls of the mobile elevated work platform. In contrast, rule-based and knowledge-based mistakes may be reduced through improved training and/or supervision.

7.4 Active Errors and Latent Conditions

Reason (1990) drew another distinction between active and latent errors. Active errors are most likely to be made by frontline workers and have an immediate effect; for example, omitting a step in a process or applying a rule incorrectly (Gordon 1998). Latent errors are removed from the 'sharp end' of work and have a delayed consequence.

Table 7.1 Illustrative example of the types of human failure.

Failure type	Example related to mobile elevated work platform operation (MEWP)
Slip	• Selecting the wrong control on the panel. • Moving the control in the wrong direction to that intended.
Lapse	• Forgetting to operate the toggle between drive and height modes on a scissor lift. • Forgetting to take account of rotation on a boom MEWP when operating drive controls.
Rule-based mistake	• Having worked previously on a MEWP and become familiar with the control configuration, a worker might not check control characteristics on a different MEWP before using it. They might operate a control that was correct for the function they wanted on the old model, but incorrect on the present one. • Familiarity with a site and route of travel can lead an operator to fail to check ground conditions, resulting in a wheel dropping down a newly cut floor recess. This could result in an overturn or significant unexpected movement of the platform.
Knowledge-based mistake	• Lack of awareness of hazards associated with using a MEWP could lead an untrained operator to perform a task in an inappropriate way; for example, near obstruction hazards, with too great a load, in windy conditions, on unstable ground. • In manoeuvring the platform close to an obstruction, an inappropriate sequence of boom movements can cause unexpectedly rapid movement, and significant 'bounce' or overrun. This could cause the operator to strike the obstruction.

Source: adapted from Leah et al. (2013).

In his later writing, Reason changed this terminology to refer, instead, to latent conditions, stating that such conditions 'arise from decisions made by designers, builders, procedure writers, and top-level management. Such decisions may be mistaken, but they need not be' (Reason 2000, p. 395). These conditions can lie dormant for long periods of time until they combine with other triggers to produce an accident opportunity.

According to Reason (2000), latent conditions produce two kinds of undesirable outcome. First, they can create the conditions in which people are more likely to make active errors – for example, by creating time pressures, fatigue, under-resourcing, or specifying the use of inappropriate equipment for a task. Second, they can produce deficiencies in system defences – for example, by providing unreliable warning systems, poorly designed facilities, or unworkable procedures.

Reason's (2000) 'Swiss cheese' model of defences, barriers, and safeguards in a work system is probably the best-known representation of how active failures and latent conditions can produce accidents. The Swiss cheese model posits that work systems have multiple defensive layers. These can rely on technologies, people, or rules and procedures. These defences are usually effective, but no defence is 100% reliable. Thus, the defensive barriers are likened to Swiss cheese, in that they have holes. Generally, these holes, which open up, close, and move all the time, do not present a problem for system safety. However, if the holes momentarily line up across multiple defensive layers, a trajectory of accident opportunity is produced. Reason explains that the holes in defensive layers are produced by active failures (errors) and latent conditions. The Swiss cheese model was reproduced in Chapter 3 in a discussion of incident causation models.

7.5   Human Error – A Cause or Symptom of Failure?

Reason (2000) suggests there are two ways of viewing human error and describes these as a person approach and a system approach. A person approach focuses attention on the unsafe acts and procedural violations of frontline workers, and sees these as arising from abnormal mental processes including forgetfulness, inattention, carelessness, and poor motivation. People who adopt this viewpoint 'treat errors as moral issues, assuming that bad things happen to bad people' (p. 393). Typical countermeasures associated with this viewpoint are designed to constrain behaviour and reduce unwanted variability. They include establishing and publicising rules/procedures, promulgating information about safety rules at workplaces, and establishing processes to retrain, punish, or even take legal action against people who break the rules.

In contrast, a system approach regards error as inevitable, acknowledging human fallibility. Errors are seen as outcomes rather than causes and are traced back to 'upstream' factors in the design and organization of work. The focus is then shifted from trying to change people to trying to create work systems with sufficiently robust multilevel defences that reduce error-provoking properties (or error traps). Reason (2000) identifies serious problems with adopting a person viewpoint on safety because many unsafe acts have logical explanations if viewed in the context of the organizational or workplace environments in which they occur. This point will be discussed in detail in the analysis of rule violations later in this chapter (see Section 7.7).

In an earlier analysis of safety in the aerospace industry, Dekker (2002) also contrasts an old view of human error with a new view. In the old view:

- human error is understood to be the cause of the majority of incidents;
- it is assumed that the system in which people are operating is intrinsically safe and, therefore, the main threat to safety comes from people and their unreliability; and
- the attainment of safety is seen to require protecting the system from humans' propensity for error through selection, proceduralization, automation, training, and discipline (Dekker 2002).

A new view of human error proposed by Dekker (2002):

- regards human error as a symptom of trouble deeper inside the system;
- does not assume systems are inherently safe – rather systems are contradictions between multiple goals that people have to pursue, juggle, and sometimes trade-off in their efforts to create safety; and
- human error is systematically and closely connected to features of people's tools, tasks, and operating environments. Understanding and addressing these connections is needed to produce safety.

Reason and Dekker both position human error as a symptom, rather than a cause, of safety failures. Dekker (2002) is particularly critical of attempts to count the frequency (and therefore importance) of human errors in safety incident causation, arguing such counts are meaningless and do nothing to explain why human errors occurred, or help us to prevent them in the future.

Dekker (2005) argues that error classification tools used in incident investigations continue to take an old view of human error, even though they may shift the focus of blame upstream to a higher level of management (or the so-called 'blunt end' of an organizational system). In doing so, such investigation tools do not reflect the true meaning of systems thinking because they fail to capture the fact that incidents/accidents are not events that occur at the endpoint of a simple linear trajectory. Rather, they are an emergent feature of complex, dynamic, and interactive social and technical processes (Dekker 2005).

7.6 Rules as a Means of Controlling Behaviour

Safety rules and procedures are fundamental components of an organizational safety management system. It is through establishing rules and procedures that managers' expectations for safe working are believed to be translated into the way work is actually done. Rules and procedures specify how a work process, task, or activity should be undertaken, and are seen to be essential in directing, standardizing, and monitoring work (Hale and Borys 2013a). Rules are seen as a means for establishing and maintaining organizational control, as a mechanism for coordination, and as a form of codified organizational knowledge (Weichbrodt 2015). However, as safety management systems have become more prevalent, safety has been criticized as being overly 'proceduralized' (see, for example, Bieder and Bourrier 2013). Weichbrodt (2015) observes how formal procedures often grow in size and volume as incident/accident investigations bring with them public pressure to take action. In these circumstances, establishing a new procedure or rule is seen as a relatively easy (and cheap) response.

LePlat (1998) observes that WHS-related rules have value only in as much as they are instruments for improving safety. However, research is beginning to question the usefulness of overly prescriptive rules or procedures, particularly when they are written by technical specialists or managers who may not fully understand the situational contingencies in which tasks are performed.

It is generally acknowledged that procedures cannot cover all eventualities and, by necessity, workers use their judgement and experience to continuously transform rules into practice. Dekker (2003) describes how safety is not always achieved through rote rule following. Rather, safety results from people being skilful at judging how to apply rules to particular contexts or situations. In some instances adaptation may be good for safety. However, this presents a 'double bind' because although rules may need to be adapted in some situations, some adaptations can also fail with serious consequences (Dekker 2003). Dekker (2003) explains that to properly understand informal systems of getting work done, it is important to continually monitor gaps between procedures and practices. Where these are not safe, the fundamental issues of work design and equipment can be addressed.

Iszatt-White (2007) identifies how formal safety documents are developed at a general level of abstraction, often by people who may not understand the applicability of rules or procedures to local conditions. Informal ways of working are important but are not always well understood or acknowledged, despite being of great importance to the way workers stay safe (Gherardi and Nicolini 2002). Ozmec et al. (2015) describe how, in small business construction firms, work practices are constantly negotiated as workers draw on personal experience and emotions, and constantly juggle safety issues with other considerations relating to workflow, customer satisfaction, and good work relations with supervisors and co-workers.

Expecting unthinking compliance with rules can also be detrimental, as workers are reduced to 'robots' unable to understand and analyse situational risks and respond appropriately to environmental contingencies (Hollnagel 2014; Fucks and Dien 2013). There is a growing belief that variability in human performance is inevitable, and reflects the potential for people to exercise skill in determining how rules or procedures should be applied in a given situation (Knudsen 2009). Resilience engineering is focused on valuing human performance variability and learning from safety successes, not just failures (Hollnagel 2011).

In some cases, excessive documentation of WHS activities in the construction industry has been perceived to distract managers' and workers' attention from more important aspects of WHS. In other industries, onerous paperwork associated with WHS management has been found to reduce managers' availability for hands on supervision of work (Lamvik et al. 2009).

7.7 Rule Violations

Hudson et al. (1998) argue that because many of the controls put in place for WHS risks are administrative, the operation of a WHS management system is based upon an assumption that people will follow procedures and rules. Consequently, when this assumption is broken, the whole basis of the WHS management system is jeopardized.

Rule violations are distinguished from other forms of error. Lawton (1998, p. 78) defines safety-related rule violations as 'deliberate departures from rules that describe the safe or approved methods of performing a particular task or job'. Safety-related rule violations contribute to accidents and safety issues in many industries, including building (Mason et al. 1995; Baiche et al. 2006).

Lawton (1998) differentiates between acts committed with the intention to cause harm (for example, acts of sabotage or terrorism) and those not intended to cause harm. Reason (2013) further classifies violations as follows:

(i) routine violations, committed to cut corners, avoid unnecessary effort, or bypass unworkable procedures;
(ii) thrill-seeking/optimizing violations, committed to make tasks more exciting or rewarding;
(iii) necessary violations, committed just so that a job can be completed; and
(iv) exceptional violations, committed during extreme one-off events when a system is operating outside normal parameters.

Alper and Karsh (2009) also suggest violations can be committed unintentionally. However, this contradicts Lawton's definition, which includes an element of behavioural intent. Unintentional violations may be better referred to as rule-based mistakes (see Figure 7.1).

7.8 Why Do People Break the Rules?

Given their prevalence and potential for serious consequences, it is important that the causes of violations be understood. Alper and Karsh (2009) suggest violations have not been more carefully studied because they are assumed to be actions taken by deviant workers. In reality, the causes of violations are complex and include individual, as well as environmental and system problems (von der Heyde et al. 2015). Hudson et al. (1998) suggest most violations result from well-intentioned workers simply trying to get their work done. Indeed, whether something is regarded as a violation or not may depend upon the outcome. Thus, a deliberate failure to follow a procedure may be seen as a violation if it goes wrong, but if nothing goes wrong and work is successfully completed, the same action may be seen as an example of a worker showing initiative (Hudson et al. 1998).

Not all violations take the same form or share the same causes. The study of rule violations has revealed a broad range of contributing factors (English and Branaghan 2012). For example, Alper and Karsh (2009) report that individual motivation, work system/organizational factors, and aspects of the external environment, interact to produce rule violations. Nordlöf et al. (2015) report that workers' risk-taking behaviour can be traced to various social and technical risk factors. Nielsen et al. (2015) provide evidence that workers with high male role norms are more likely to violate safety rules and less likely to report violations, causing particular concerns for male-dominated industries such as construction.

Violation is a pejorative word that suggests deviance and wrongdoing (Alper and Karsh 2009). However, as Dekker (2005) points out, organizational environments often drive multiple competing goals. For example, in a construction project, emphasis is placed on cost, time, and quality performance, as well as WHS. Dekker describes how these objectives are internalized by members of an organization (the same would be true of project teams), such that workers pride themselves on their ability to make daily adjustments and trade-offs to manage these competing demands. In this context, shortcuts can become routine. What seems like a clear case of deviant behaviour from the outside (usually with the benefit of hindsight) can look like a perfectly reasonable action to the person who took it.

It is also important to identify when safety rules and procedures are ill-suited to particular work tasks, situations, or contexts. In these instances, the rules themselves may need improvement (Pilbeam et al. 2016).

Hudson et al. (1998) suggest no rules/procedures are perfect. Violations vary by type in relation to limitations inherent in rules or procedures established in a particular context. Thus:

- routine violations occur when poor or unworkable procedures are not followed;
- situational violations occur in special situations that are poorly covered in procedures; and

- exceptional violations occur when situations are not covered in procedures (Hudson et al. 1998).

Case Examples 7.1–7.3 identify occurrences of workers' violations of WHS-related rules in the Australian construction industry and the circumstances in which violations occurred. These examples of rule violations were observed during a study that involved using participatory video. During the study, workers made films about the WHS aspects of their work practices.

Case Example 7.1 Inadvertent Rule Breaking in the Australian Building Industry

Several instances of unintentional rule breaking were observed among a team of insulation installation workers. Some safety-related rules were routinely broken because workers did not know the rules existed. For example, workers were unaware that their standard operating procedure for installing insulation in wall cavities required them to maintain 600 mm clearance when using a nail gun to fix insulation adjacent to an electricity conduit. One worker commented: 'I heard it when we did an audit recently … but with the conduit and stuff like that, I haven't really steered 600 mm clear of that'. Another commented: 'I actually didn't know about shooting all the pins 600 mm either side of the conduit. That's one thing that I did not know personally … Yeah, like I said I thought it would have been something that someone would have told me in my whole learning process of being here, but obviously it doesn't seem to be happening.'

Although the 600 mm clearance requirement was documented in the standard operating procedure, workers described how they did not read safety documentation in detail. One said: 'A lot of the times you just skim over it and you sign the back half the time. As bad as that may sound, everyone does it … How often is someone going to actually sit down and read every little detail? I know we should and everyone should, but when it comes down to it, no-one really hardly ever does.'

Source: Lingard et al. 2016.

Case Example 7.1 describes an example of inadvertent rule violation (or a rule-based mistake). Case Example 7.2 describes an example of a routine rule violation workers made to bypass an unworkable procedure. Case Example 7.3 describes a situation in which commercial building companies applied pressure to subcontractors to violate important WHS rules, with the potential to disrupt the work of other trades.

LePlat (1998) identifies accessibility and legibility of WHS rules as critical determinants of their implementation. In particular, it is important that workers can find and comprehend rules. The mode of presentation is likely to be a key factor in accessibility and legibility. In the case of the insulation workers, presenting rules in written form was problematic. The workers identified low levels of literacy as a reason why they did not want to read procedures. One commented: 'There's just so much information and it's just not practical to sit there for three, four hours because I'm not very good with the English language. So for me to read a document like that would take me half a day and they're not going to let you sit there and do that … You're also going to embarrass yourself in a room with 20 other people … So there's pressures to sign them off.'

However, in the above quote, the worker also reveals how commercial pressures experienced by subcontractors 'to get the job done' interfere with the implementation and effectiveness of safety rules. This reflects what Iszatt-White (2007) describes as a 'gambit of compliance', in which a box is ticked to say workers have read the standard operating procedure, but no-one actually checks whether rules are properly understood and followed. In effect, some building companies 'turn a blind eye' in the interests of getting the work done.

Case Example 7.2 Routine Rule Breaking in the Australian Building Industry

A different group of insulation installation workers was filming the task of insulation installation in ceiling spaces in domestic buildings. It became apparent to them that it was practically impossible to follow the standard operating procedure for the task.

The work involved accessing ceiling manholes at a height of between 2.4 and 2.7 m from the floor. The company's standard operating procedure for using ladders and working at height requires that a straight ladder be placed at a 1:4 ratio and extend 900 mm beyond the 'step off' point. A script for the film was developed and distributed for comment. The health and safety manager commented that 'no-one had an issue with [the script] theoretically'.

However, on the day of the filming, the worker who was to undertake the task was furious, arguing the script did not reflect the way the task is routinely undertaken. The health and safety manager describes how 'shooting it [the film] and viewing it through the camera's eye, we had to stop … the camera doesn't lie'.

She explained: 'To place a straight ladder at the 1:4 ratio just doesn't work. You can't get a body in there as well because it blocks off the access and you have to contort yourself to actually get in [to the ceiling space]'. The requirement that the ladder extend 900 mm beyond the 'step off' point was physically impossible to achieve due to conduits, cables, beams, and other obstructions. The small size of the manholes did not allow adequate entry for the ladder, the worker, and the pack of insulation to be installed. The health and safety manager described how workers passing insulation packs into the ceiling space using a straight ladder had to contort their bodies to manoeuvre themselves into the ceiling space then move the ladder to get the packs in. She also explained that, if the workers used an A-frame ladder, 'which they do because they can't use a straight ladder', they are forced to work unsafely because they have to step off the top rung of the A-frame. This practice is also in breach of the standard operating procedure for using ladders and working at height.

Source: Lingard et al. 2016.

Case Example 7.2 reveals a gap between the way work was documented in the standard operating procedures, and the way it was routinely performed. This gap became apparent to the health and safety manager when making a film depicting this task (see also Hollnagel 2015). Workers had long experienced this gap between work as imagined and work as done. Hudson et al. (1998) describe how routine violations often occur with such frequency that they become automatic and unconscious behaviours. In this way,

---

Case Example 7.3 Expectation of Rule Breaking in the Australian Building Industry

The insulation installation workers also described being asked and expected to violate their company standard operating procedure by the general builders (principal contractors) who engage their services. In some instances, this led them to take potentially life-threatening risks.

The insulation workers made a film showing the practice of isolating and 'locking out' the electricity supply before commencing installation work in buildings at which the electricity supply was live. The construction sites at which the installers work are typically under the control of a builder (principal contractor). The installers described how they often receive no information from builders about the location of cables in walls and ceilings, or about whether any cables are live. One reflected: 'If we're shooting into concrete with live wire, then you would think someone would say something, or show you a plan.'

The workers also discussed the resistance they face from builders who do not want to isolate the electricity supply, even temporarily, so that the installation workers can work safely. The health and safety manager described how the insulation firm had been threatened with termination of contract when a builder learned about the company's electrical isolation procedure. She describes how 'a situation had developed out onsite where it had become a gentleman's agreement [sic] between the trades and ourselves: "okay you need power so we won't isolate because of the inconvenience that it would cause out on job sites." Rather than challenge the status quo, that situation had become expected.'

This example shows how managers at the insulation firm were trying to look after their workers by implementing an electrical isolation procedure. However, there is an expectation among the builders who engage the insulation firm that safety-related rules will be bent or broken to get the job done.

Source: Lingard et al. 2016.

---

routine violations are often seen as presenting little risk and have become the normal way of performing a particular task. Group acceptance of the violation means it is unlikely to be remarked upon or corrected in everyday work conditions.

Hudson et al. (1998) describe how situational violations occur as a result of factors dictated by workers' immediate work environment, which make it difficult for them not to commit a violation. This could be due to time pressure, lack of available equipment, or low staffing levels. Pressures to complete work and maintain production have been found to reduce compliance with safety-related rules in construction and other industries (Guo et al. 2016; Dahl 2013). The insulation installation workers were positioned at the lowest level in the building industry's hierarchical supply chain. Case Example 7.3 reveals that, even when they know about the safety-related rules that apply to their work, the insulation workers are expected to break these rules by the general builders who engage them. The expectation that workers break rules to get the work done reflects power relations between companies in which economic and reward pressures become successively greater towards the bottom of the supply chain. The problem is exacerbated because subcontracted workers often work in small firms, may not represented by trade unions, and experience job insecurity and precarious employment (Quinlan 2011). The practice of shooting nail guns into walls that possibly contain live electricity cabling is potentially fatal. See Maiden 2010 on the risks of electrocution when installing insulation. Yet this risky practice is tacitly accepted (even expected) by building companies.

7.9   The Importance of Rule Management

The case examples above illustrate how construction workers break WHS-related rules for many reasons. In workplaces the focus is often on enforcement of rules, rather than a critical examination of the work environment or the suitability or operation of rules themselves.

Hudson et al. (1998) suggest the following questions should be considered as precursors to designing strategies to improve compliance with rules/procedures:

- Do employees know and understand the procedures?
- Do we need all of these procedures?
- Are there situations when it is impossible to apply procedures?
- Does the job itself encourage violations?
- Is it possible to have a procedure for every situation?
- Are there alternatives to procedures?

Hale and Borys (2013b) recommend making rule monitoring and improvement an explicit and central activity in managing rules and procedures in a workplace. They also recommend that people who undertake the work tasks for which rules/procedures are being developed participate in rule making and monitoring. Rather than focusing solely on developing and communicating rules, rule management also needs to include processes for monitoring and changing rules when necessary.

The process Hale and Borys (2013b) recommend for managing rules is depicted in Figure 7.2. This represents a cyclical structure in which rules are adapted as necessary to the changing realities of work processes and/or the work environment. The model assumes an organization starts with some existing work procedures and rules, but the first and second steps in the process involve monitoring individual or group use of rules, and seeking feedback about rule use and effectiveness. This then leads to a cycle of evaluation, enforcement of good rules, and developing ways to deal with exceptions (that is, situations in which the rules do not apply). Having identified exceptions, Hale and Borys (2013b) suggest mechanisms for coping with exceptions are required which must match workplace culture and workers' capabilities to cope. These steps (1–4 in the model) provide an opportunity to monitor gaps between procedures and practices and allow rules to be adapted as necessary to suit local conditions.

Hale and Borys (2013b) suggest tacit knowledge derived from practice should play a key role in evaluating rules and procedures, but also recognize the need for this knowledge to be made explicit and subject to peer or technical review.

Having evaluated rule effectiveness and understood the reasons for violations, errors, and exceptions, the model proposed by Hale and Borys (2013b) suggests bad or superfluous rules be scrapped or redesigned (step 5). They suggest all rules may

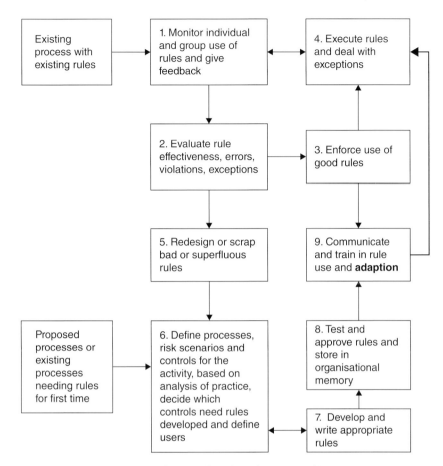

Figure 7.2 Rule management framework (Hale and Borys 2013b)

even have a 'lifespan' and that rules may usefully be subject to periodic automatic review. In revising procedures/rules, a task risk analysis (step 6) is advocated. This stage provides an opportunity for workers and their health and safety representatives, managers, and others to discuss risks and controls, and consider the role played by WHS rules. Designing rules appropriate to particular audiences is a particular challenge as auditors and regulators may favour long and detailed documents, while frontline workers may find such documents difficult to read or understand. Hale and Borys (2013b) recommend that the role of rules/procedures is clearly explained and taught to rule users (and their supervisors) so that people fully understand how much discretion they have in following rules.

Step 7 in the model involves developing and writing appropriate rules. Hale and Borys (2013b) suggest the form these rules take should be appropriate to the level of workforce competence and experience. Whether procedures are viewed as prescriptive action rules, or resources for action, has been identified as a key difference in what procedures mean and the way they are operationalized (Dekker 2005). Dekker

(2005) describes applying procedures as a substantive cognitive activity in which rule users exercise their judgement and adapt procedures to fit the circumstances they face. For example, Carim et al. (2016) show how procedures are used to guide decision making and action of flight deck crews in commercial airlines. Given the complicated context, pilots employ strategies that draw upon multiple resources to solve problems, and often follow small parts of documented checklists, rather than following the whole procedure from start to finish. Thus, when the workforce is highly skilled and competent, and the work context is complicated, prescriptive action rules may be less appropriate.

However, rules also need to be designed to provide a level of practical detail appropriate for specific work tasks, taking into account workforce capability. In a study of Safe Work Method Statements in the Australian construction industry, Borys (2012b) reports that supervisors play a key role in communicating to workers the informal adjustments that can be made to formal action rules to suit local circumstances. Thus, supervisors' competence in making and communicating 'on the spot' adaptations may be particularly important in the construction context.

Hale and Borys (2013b) also recommend a trial use of proposed rules to ensure they are workable and internally consistent (step 8). Criteria for evaluating the performance of newly developed rules vary but could include the extent to which they are understandable, reproducible, clear, and valid (evidence-based). Finally, Hale and Borys (2013b) propose training (step 9) to ensure people know about the new rules and, importantly, how and when to apply the rules. Where rules are intended to be resources for action, users should be provided with the requisite knowledge and understanding to be able to adapt the rules to the circumstances they are likely to experience.

Hale and Borys (2013b) argue their proposed model for a rule management process can reconcile contradictions between top-down models that emphasize formal procedures and action rules that prescribe behaviour, and bottom-up models that emphasize adaptation and flexible deployment of tacit knowledge. Their model does this by acknowledging and managing exceptions, engaging rule users in evaluating and redesigning procedures, and allowing for adaptation and flexibility. At the same time, the rule management model reflects the important role played by procedures and rules in maintaining WHS, and providing transparency in the development of procedures/rules appropriate to a particular industry context or organization.

Construction workers present a largely untapped source of WHS information because traditional WHS approaches have focussed on top-down management control of workers (Saksvik and Quinlan 2003). Such traditional approaches emphasize enforcing workers' compliance with safety rules established by managers and technical specialists. There is evidence that organizations with mature cultures and high WHS standards actively seek employee participation and feedback in WHS planning, decision making, and improvement (Health and Safety Executive 2005a; Törner and Pousette 2009). Yet Ayers et al. (2013) describe how, despite a statutory requirement to consult workers, most construction organizations do not involve workers in making strategic WHS decisions. For example, it is rare for workers to be asked to participate in designing workplaces or systems of work. Frick (2011) similarly suggests workers' ability to influence WHS in a voluntary WHS management system is determined by an organization's intentions in implementing the management system. If the management system's deployment is motivated by internal desire to improve performance and workers' health, safety, and welfare, then meaningful consultation with the workforce will be needed. However, if the management system's deployment is motivated by external goals (for example, to improve a company's brand image, or manage injury statistics for external reporting), then consultation will be ineffective, perhaps limited to a one-way (top-down) communication of safety rules.

This is a missed opportunity because workers possess a wealth of knowledge about WHS hazards associated with construction tasks and about ways to work more safely. Much of this knowledge is tacit; that is, it is difficult to transfer to another person by means of writing it down or verbalizing it (Polanyi 1958). This type of knowledge can be described as 'know how', rather than 'know what'. For example, knowing how to use a complex piece of equipment or perform a complicated work task safely are forms of tacit knowledge. In many cases the people who possess tacit knowledge are not aware of their knowledge and do not know how valuable it could be to others. Construction workers may not even appreciate the extent and value of their WHS knowledge and may be unlikely to possess the skills to easily communicate their knowledge to others. Tools that can help to unlock the tacit knowledge of workers have significant potential to improve WHS.

Worker engagement describes how workers can be encouraged and supported to take part in decision making about health and safety management (Meldrum et al. 2009). Research has demonstrated the benefits associated with engaging workers in WHS improvement processes. Wachter and Yorio (2014) show how workers' cognitive and emotional engagement in WHS mediates the impact of an organizational WHS management system on WHS performance. Thus, it is through the ability to engage the 'hearts and minds' of workers that formal WHS management processes achieve their best results. Indeed, a participatory (worker-led) approach to developing WHS capability has been found to produce demonstrably better results in terms of knowledge acquisition and injury prevention than traditional WHS training approaches (Burke et al. 2006).

Using participatory video in the Australian construction industry, Lingard et al. (2015b) were able to access workers' tacit knowledge of WHS to inform meaningful improvements to WHS procedures and practices. These are documented in Case Example 7.4 below.

Participatory video is a 'group-based activity that develops participants' abilities by involving them in using video equipment creatively to record themselves and the world around them and to produce their own videos' (Shaw and Robertson 1997, p. 1). Unlike observational cinema, participatory video is reflexive, and in our case the workers were both subjects and film makers.

Workers were engaged in making films about the safety aspects of their work. On completion the worker-made films would be shared with other workers as safety training resources. The participatory video process was facilitated by an external consultant who engaged workers in brainstorming content, developing a story board, and filming

Case Example 7.4 Unlocking Workers' Tacit Knowledge Using Participatory Video

Filming and reviewing film footage of workers undertaking their everyday work tasks enabled a number of previously unrecognized WHS issues to be identified and resolved.

In the process of filming and producing a video depicting the erection of a mobile tower scaffold, it became apparent that workers were exposed to a fall hazard. This hazard was present in the standard erection process for the tower scaffolds but, because the period of time for which workers were at risk was relatively brief, the hazard had been accepted as part of the erection process. The issue was identified during the filming of workers performing this erection task and this enabled a solution to be found and recorded. This improvement opportunity would have been difficult to identify by referring to documented erection procedures. Probably it would not have been identified were it not for the opportunity to watch and analyse the video recording of the erection process.

The scaffold supplier and work crew were involved in making the film. They filmed the construction method for the mobile scaffold and took the video to the site office to review it. While watching and reviewing the footage, the project team identified a period of time during which workers erecting the scaffold tower had no fall protection. As a manager described it:

> There was just one phase, for 30 seconds, where they were unprotected and I said, "I'm sure we can do something different" … So we went back out to the worksite with the crew, the supplier and he showed the crew the issues and said, "How do you reckon we fix it?"

The crew spent several hours trying different erection processes and procedures and eventually worked out a new method for erecting the scaffold tower without having the window of exposure to the risk of falling from an unprotected edge. The manager commented:

> The previous way of building [the scaffold] had been custom and practice for decades … no-one had sort of thought twice about it, but once you saw it on the screen it didn't look quite right … And we just got the guys who had been doing it for years to try and find a way to fix it, and in the end they did.

The revised sequence used temporary mid-level platforms and horizontal rails to ensure workers could work within in a side structure at all times during the erection of the scaffold.

The identification of safety issues inherent in organizations' standard operating procedures was a recurrent theme in the data collected from the interviews. In another situation, workers identified a safer access system after reviewing video footage of a hazardous work task undertaken from a barge in an aerated sewerage channel.

Workers were coating concrete sewerage channels with epoxy to prevent corrosion. The task was very dangerous because the channels contain aerated liquid. A fatality had occurred six months prior to the operation at another facility when a worker fell through

Case Example 7.4 (Continued)

a dislodged cover into an aerated channel and drowned. The prosecutor in that case commented that if someone were to fall into a channel like this, it is almost certain they would drown.

To apply the epoxy to the channel beams, work was performed from a barge. The operation for accessing the barge safely was filmed, and the film was reviewed by workers and managers at the site.

The project manager described how in 'even just a little thing, how they hooked themselves onto the safety line we identified there was a little gap. Like, there was virtually a 10 second gap [during which time] they weren't hooked on'. The project manager described how the problem inherent in the process of entering the barge became apparent, noting it was 'only when they [the workers] acted it out that they were conscious of "hang on, because you have to actually undo [the harness] from the side and attach it to the base of the boat"'. There was a period of time during which the workers had no protection against falling into the sewage channel. As a result of this observation, a new work process was developed in which multiple connection points were used. This meant workers were protected from falling into the channel at all times. The project manager commented: 'So it was only with just acting it out and the crews themselves identified it.'

Workers who participated in the participatory video process were enthusiastic about the visual representation of work activities and felt that film was a much better way to understand and learn about safer ways of working. For example, one said: 'You see people doing it, see what has to be done or [what] you shouldn't be doing so it's better than reading.' Workers were also much more likely to engage with, and critically review, rules and procedures when they were presented as film. One commented:

> It's a lot easier to show someone what we're trying to say. We could just sit here and verbally speak about it but if you put your verbal words into a video, people are going to sit back and go, 'now I know what he's actually trying to say'.

He continued:

> See how I can talk about stuff, this and that, and in your own head you'd get your own visual perception of what's meant to be going on, rather than someone actually going there, showing you, going, 'okay look, I reckon it's definitely a better way of getting your point across … visually showing someone'.

Workers believed written rules and procedures could never convey the way work is actually performed. One explained:

> We can't put everything in there [the WHS procedures] because sometimes it's something you can't write in there because it's knowing … You can make a note but, yeah, it's hard to describe everything in words.

(Continued)

Case Example 7.4 (Continued)

Managers also observed that workers were actively engaged in identifying gaps between work as imagined (in written procedures) and work as performed. One participant commented:

> During the process, it's amazing how many things the crews themselves picked up … They said: 'Ooh, actually what we do in practice is not what the document says. Actually we do this.' So it actually I think resulted in improvements to written safe work method statements to better reflect what crews were doing.

Workers indicated they enjoyed the participatory video process, in particular in helping to shape improvement of rules and developing visual procedures that would be shared with others. One commented: 'It's good to give people input, especially when you realise that it's going to improve something.' Another told us: 'Yeah, it's good for everyone to throw their input in and you just learn a lot more about what could be done and it gets your mind thinking.'

The workers described how they felt valued because the company had taken the time to really understand (from their perspective) the way they work and the WHS aspects of their jobs.

For example, one commented:

> It was actually good, you know, getting my voice heard and actually people sitting there and actually taking it in what I was saying, rather than just going over someone's head or they just turning a blind eye to it. It actually feels like something or progress can be made or people actually listened.

The workers also commented that the content of the resulting visual procedures was likely to be more appropriate and useful because workers with experience were consulted and allowed to have input into the procedures depicted. One worker commented: 'Yes, I think it's the best way because if [the material] is from people working on the site, it's going to be spot on. Exactly what is happening.'

Another worker who has extensive construction industry experience in another country also indicated he was very happy to be able to share his knowledge, commenting: 'I like they asked me because I can share what I know. It's good.'

Consistent with this positive response from workers, a WHS manager described how workers were actively engaged in making suggestions about WHS process improvements during the filming: 'The camaraderie is great, which opens up free thinking and free speech and things come forward. When you ask their [the workers'] opinion, they feel valued.' Another manager explained: 'They're not afraid to bring forward their ideas because they think everything's going to be considered.'

In particular, workers' involvement in the participatory video process provided them with an opportunity to improve and contribute to the design of work processes. One WHS manager described workers' antipathy to written WHS procedure documents that are often developed without workers' input. She contrasted this with participatory

---

Case Example 7.4 (Continued)

development of visual WHS procedures, saying: 'They [the workers] can see that it's not fixed in concrete. They have an ability to improve it and contribute to it and it's made up of what comes from them.'

Such was the effectiveness of the process that one manager described how 'pretty soon the guys started coming to us with ideas about more stuff we could do'.

Source: adapted from Lingard et al. 2015d.

---

their work. Participatory video is more than a data collection approach; it is, in part, also an intervention. The aim is not simply to produce video materials, but to use the process of video production to generate critical thinking and 'empower people with the confidence, skills and information they need to tackle their own issues' (Shaw and Robertson 1997, p. 26). During the making of the film, a member of the research team observed the film making and undertook video-recorded interviews exploring work practices.

7.10   Conclusions

In many organizations WHS management systems have become heavily proceduralized, relying on the establishment of formal, written-down rules as a means to control behaviour and ensure standardized practices and performance. Highly bureaucratic management systems have been criticized for producing overly long, complex, and sometimes unhelpful procedures prescribing the way work must be undertaken. Such an approach seeks to control and coordinate behaviour and create a shared body of knowledge about how work should be performed in a particular organization or workplace.

However, the usefulness of top-down control of work to reduce variability in human performance has been questioned, not least because human variability may not, in fact, be the problem. The attribution of safety incidents/accidents to human error is very common, and is reinforced by investigation procedures and human error classification systems. However, the term 'human error' is often used uncritically and inconsistently. Sometimes errors are framed as causes, other times as deviations from standards, or the failure itself. Depending on which view of error is taken, the usefulness of prescriptive behavioural controls may have limited influence in preventing errors.

Errors can be seen as symptoms of deeper problems in systems of production, related to conflicting goals, organizational pressures, and the design of work itself. Understood in this way, the development of rules, the training of workers, and exhortations to 'take more care' are unlikely to significantly reduce human errors because more fundamental causes of errors (and WHS performance) are at play. Heavily documented WHS management systems are also limited in the extent to which they can prevent intentional violations of rules/procedures. Violations occur for many different reasons, but in many cases rules are violated by well-intentioned workers simply trying to get their jobs done. Evidence from the construction industry suggests that, in some cases, subcontracted workers are pressured to violate rules to complete work with minimal disruption. Also, violations can occur when there is a mismatch between prescriptions in formal procedures and the physical reality of a work environment, or when workers do not know about or haven't properly understood a procedure. Workers' participation in evaluating, designing, and testing procedures is likely to improve their effectiveness and, ultimately, compliance.

A participatory rule management system is therefore recommended, by which rules/procedures are systematically reviewed and revised. Given construction workers' preferences for visual communication and learning, the use of participatory video as a mechanism for the review and redesign of WHS procedures has considerable potential. Further, disseminating visual procedures depicting practical information about safe ways of working is now possible with the use of digital mobile technologies, such as smartphones.

Discussion and Review Questions

1 How useful is the term 'human error' in investigating and identifying the causes of WHS incidents/accidents?

2 Why do people sometimes break rules related to WHS?

3 How important are prescriptive action rules in maintaining WHS in construction? How can rules be balanced with the need for adaptation to local conditions?

Acknowledgements

The participatory video research received support from the Victorian Government Department of Business and Innovation through its Business Research and Development Voucher program. The work was also part-funded and supported by CodeSafe Solutions.

8

An Integrated Approach to Reducing the Risk of Work-Related Musculoskeletal Disorders

Helen Lingard

School of Property, Construction and Project Management, RMIT University, Melbourne, Victoria, Australia

8.1   Introduction

Throughout this book, an argument is made to integrate work health and safety (WHS) more fully into construction project management processes. Chapter 2 discussed the role of clients in establishing high expectations for WHS at the outset of a construction project and in using the procurement process to create project conditions supportive of WHS. Chapter 3 presented the arguments for considering and addressing WHS issues during the design stage of a construction project and, in particular, paying attention to the design of the construction process as well as the product (that is, building or other facility). Chapter 4 described occupational health risks experienced by construction workers and discussed the need to tackle the issue of construction workers' health by considering the quality of jobs and employment in the sector. Chapter 5 addressed the role played by project and organizational cultures on WHS and identified a number of workplace culture characteristics that will enable (rather than impede) WHS. Chapter 6 addressed the question of how to measure WHS and argued for the use of alternative measures to evaluate risk reduction efforts, and to understand workplace cultures and their WHS impacts. Chapter 7 discussed human behaviour and considered the reasons why people sometimes break WHS-related rules. The opportunities afforded by engaging workers in the design of work processes and procedures were discussed.

This chapter ties all these arguments together in an in-depth discussion of work-related musculoskeletal disorders (WMSDs) in the construction context. Previous chapters have not targeted specific occupational hazards, injury, or illness types, but have presented general ideas or arguments about how WHS could be improved in the construction industry. In contrast, this chapter aims to show how these ideas and arguments can provide insights about reducing WMSDs in construction.

A general overview of WMSDs in construction is provided, describing the size and apparently intractable nature of the problem. Factors contributing to WMSDs in construction are then considered, and linked back to aspects of the work environment and project cultures shaped by clients, principal contractors, and others. Opportunities are

Integrating Work Health and Safety into Construction Project Management, First Edition.
Helen Lingard and Ron Wakefield.
© 2019 John Wiley & Sons Ltd. Published 2019 by John Wiley & Sons Ltd.

identified for reducing the risk of WMSDs by considering particular issues at the design stage. The chapter discusses how to measure WMSD risk and, more importantly, to objectively evaluate risk reduction strategies. Finally, the chapter explores the potential for engaging construction workers in redesigning work processes by using a participatory ergonomics (PE) approach. The opportunities this approach affords for WMSD risk reduction are considered.

8.2 The Prevalence of WMSDs in Construction

Musculoskeletal disorders include 'a wide range of inflammatory and degenerative conditions affecting the muscles, tendons, ligaments, joints, peripheral nerves and supporting blood vessels' (Deeney and O'Sullivan 2009, p. 239). The US Bureau of Labor Statistics defines musculoskeletal disorders to include:

> ... cases where the nature of the injury or illness is pinched nerve; herniated disc; meniscus tear; sprains, strains, tears; hernia (traumatic and non-traumatic); pain, swelling, and numbness; carpal or tarsal tunnel syndrome; Raynaud's syndrome or phenomenon; musculoskeletal system and connective tissue diseases and disorders, when the event or exposure leading to the injury or illness is overexertion and bodily reaction, unspecified; overexertion involving outside sources; repetitive motion involving microtasks; other and multiple exertions or bodily reactions; and rubbed, abraded, or jarred by vibration *(BLS 2017).*

Construction workers are a high-risk group for WMSDs (Schneider 2001). International research suggests this problem exists all over the world. For example, in the USA, the Center for the Protection of Workers' Rights (CPWR) reports that the rate and number of WMSDs in construction fell between 2007 and 2010. Notwithstanding this, in 2010, the rate of WMSDs in construction was still 16% higher than the rate of WMSDs among full time equivalent workers for all industries. It is also important to note that these numbers may be underestimated because musculoskeletal disorders may sometimes be unreported or not attributed to workplace exposures (CPWR 2013; Dale et al. 2015).

The USA data indicates that WMSDs in the construction industry most frequently affect the back (45% of cases). The shoulders and extremities each accounted for about 10% of WMSD cases reported by construction workers (CPWR 2013).

One of the most commonly cited factors in WMSDs in construction is overexertion. The CPWR reports that, in 2010, overexertion in lifting caused 38% of WMSDs among construction workers. Other types of overexertion, including pushing, pulling, and carrying, caused a further 35% of WMSDs in construction. Some trades, including masonry and concreting, are particularly susceptible with rates of overexertion injury higher than general construction industry rates. Many overexertion injuries are sprains, strains, and tears that can develop into chronic injury and prevent people from working (CPWR 2013).

In Australia, the situation is similar. Around 12 600 workers' compensation claims are accepted from the construction industry each year for injuries and diseases involving one or more weeks off work. This equates to 35 serious claims each day. A serious claim is a workers' compensation claim for an incapacity that results in a total absence from work of one working week or more. Body stressing is the most common type of injury, accounting for 37% of serious claims. Back injuries also account for the largest portion (20%) of serious compensation claims in the Australian construction industry (Safe Work Australia 2017).

WMSDs are reported to have serious consequences in terms of sickness absence, cost (Rinder et al. 2008), long-term work ability, and workforce participation (Welch et al. 2009, 2010). In a study of Swedish construction workers, musculoskeletal disorders were the most common cause of work disability in all construction occupations (Stattin and Järvholm 2005).

There is evidence to suggest symptoms of WMSD are experienced by construction workers very early in their careers. For example, Merlino et al. (2003) studied the experiences of construction apprentices with an average age of 27.7 years. In a 12-month period, the majority of apprentices (76.8%) reported experiencing WMSD symptoms. Parts of the body most frequently affected were:

- the lower back (54.4% of apprentices);
- the wrist/hand (42.4% of apprentices); and
- the knee (38.4% of apprentices).

Similarly, in a study examining WMSDs in floor layers, Dale et al. (2015) report floor layers have higher rates of WMSDs than general construction industry workers for all body regions considered. However, the greatest difference between incidence rates occurs in the youngest group of workers (aged 18–24 years). The comparatively large incidence of musculoskeletal symptoms in young workers engaged in floor laying suggests that, even though young workers have good levels of physical strength and flexibility, when their work tasks exceed their physical capabilities they are highly susceptible to WMSDs (Dale et al. 2015).

8.3   Risk Factors for WMSDs

Risk factors for WMSDs fall into two categories:

- those related to the physical workload – for example, high static loads, handling of heavy materials, awkward body postures, and vibration; and
- those related to the psychosocial work environment (Hollmann et al. 2001).

However, these should not be regarded as independent of one another because, as Hollmann et al. (2001) point out, poor psychosocial conditions in many jobs are also associated with high physical workloads. This point is also made by Huang et al. (2003) who argue that the effects of physical and psychosocial stressors on occupational health may be traced back to common causes associated with the quality of work and how it is organized and performed.

Many studies of WMSDs in construction focus on physical risk factors. These are summarized in Table 8.1.

Physical risk factors for WMSDs vary by occupation or trade and are related to the work tasks these trades typically perform. Some specific risk factors associated with

Table 8.1 Physical risk factors for WMSD in common construction tasks.

Factor	Definition	Damage and symptoms
Repetition	Using the same muscles repeatedly without rest	Strain in tendons and muscle groups involved in direct repetition motions
Force	The physical effort required to perform a task or maintain control of tools	Stress on the muscles, tendons, and joints which is associated with risk of injury at the shoulder, neck, lower back, wrist, etc.
Awkward posture	When any joint of the body bends or twists excessively, or any muscles stretch beyond a comfortable range of motion	Sprain and strain in wrist, shoulder, neck, and lower back
Vibration	Any movement that a body makes about a fixed point	Damage caused to body organs buffered by relatively low frequency and breakdown of body tissues resulting from continued absorption of high-energy vibration
Contact stress	Injury by hard, sharp objects when grasping	Nerves and tissues beneath the skin of the wrist, palm, or fingers injured by pressure when a hard or sharp object comes into contact with the skin

Source: Jaffar et al. (2011).

particular tasks are briefly summarized below. This is not an exhaustive analysis but highlights differences in risk exposure and experience of WMSDs between worker groups.

8.3.1 Steel Reinforcement Fixing

Steelfixers engage in significant manual materials handling of steel rods, as well as highly repetitive activity associated with tying these rods together to form cage structures. Steelfixers also need to walk on unstable and uneven surfaces to access their work. Forde et al. (2005) analysed the prevalence of musculoskeletal symptoms in steelfixers. They report that, compared to workers engaged in other metal-working activities (such as constructing structural steel components or ornamental ironwork), steelfixers have the highest incidence of musculoskeletal symptoms in the wrist, hand, or fingers (48%), and have the lowest proportion of workers not affected by any WMSD symptoms. The age-adjusted odds ratio for workers having upper extremity symptoms was also significantly higher for steelfixers, compared to workers engaged in other metal-working activities. Buchholz et al. (2003) observed the work of 17 steelfixers performing five job tasks in the USA construction context. These were:

- ground-level reinforcement bar (rebar) construction;
- wall rebar construction;
- ventilation rebar construction;
- preparation work; and
- supervising.

Non neutral trunk postures were observed frequently (exceeding 30% of the time) and manual material handling was the most commonly observed activity (exceeding 20% of the time) for all job tasks except supervising. However, even when supervising was removed from the analysis, Buchholz et al. (2003) report significantly different risk factors for the steelfixing tasks. They conclude that steelfixing is a high-risk activity in which ergonomic improvements are needed, but that ergonomic interventions need to be based on a detailed analysis and understanding of the risk factors associated with specific steelfixing work tasks.

8.3.2 Concreting/Screeding

Concrete workers place concrete by pump or bucket, compact concrete in forms or slabs, and undertake formwork 'stripping', which involves removing forms from footings, walls, and slabs. Goldsheyder et al. (2004) report that concrete workers are exposed to a wide variety of WMSD risks, including heavy manual materials handling, repetitive and forceful exertions, awkward postures, frequent bending and twisting movements, work above shoulder or below knee level, and strenuous and fast-paced work. In an analysis of WMSD incidence among concrete workers, Goldsheyder et al. (2004) report that 77% of concrete workers experienced at least one musculoskeletal symptom in the 12 months prior to the survey. Lower back pain was the most frequently reported WMSD symptom (reported by 66% of concrete workers), followed by shoulder pain (47%) and neck pain (44%). Thirty-six percent of concrete workers perceived continuing to work while in pain to be a major problem in their occupation.

Removing extra concrete and levelling concrete to grade is referred to as concrete screeding. Screeding can be performed using different techniques. It has been reported that the most serious risks of developing WMSDs of the upper extremity and back are associated with manual and roller screeding (Albers et al. 2004). While the powered screeding technique is less risky than manual screeding, power screeding can include periods of medium/high exertion for lifting tasks. In addition, vibratory screeding equipment exposes workers to hand-arm vibration.

8.3.3 Floor Laying

Floor laying involves different tasks depending on the materials and surface of the floor being laid. Burdorf et al. (2007) examined WMSD risks inherent in installing sand-cement floors in new buildings and the levelling of floor bases. Traditionally, this work is undertaken by a team of two workers, one of whom spends most of the time outside the building shovelling sand into a mixer. An attached pump pushes the sand/cement through a hose to a room in the building where a floor layer pours the mixture over the surface, levels the floor with a board, and finishes the floor with a sander. Burdorf et al. (2007) report the internal floor layer has higher postural load than the external worker due to working in a squatted/kneeling position and being observed to have back flexion over 40°. The external worker lifts loads for about 20% of the work time, whereas the internal floor layer pushes and pulls a rake for about 28% of the total work time taken to perform the floor laying task. Fifty-six percent of workers engaged in laying floors in this way reported the presence of low-back pain in the six months prior to the study, with more than 50% seeking medical treatment and slightly less than 50% taking sick leave for this pain.

Dale et al. (2015) also analysed floor layers' experiences of WMSDs and report that the proportion of workers with at least one claim for a WMSD is significantly higher among floor layers than general construction workers. Further, this was the case for five different body locations considered (knee, neck, low back, distal arm, and shoulder). The biggest differences were observed for WMSD symptoms in the knee and neck, which were more than double for floor layers compared to general construction workers. Claims for WMSDs in multiple body locations were also more common among floor layers. Dale et al. (2015) note that workers engaged in floor laying use a great deal of force and adopt repetitive postures while kneeling on the floor to spread adhesive, lay ceramic tiles, and nail boards.

8.3.4 Mechanical and Electrical System Installation

Mechanical and electrical system installation includes installing service piping and heating, ventilation, and air conditioning (HVAC) systems. These activities involve different trades, including plumbers, pipe fitters, sprinkler fitters, mechanics, and electricians. Albers et al. (2005) identify service installation work as a high-risk activity for WMSDs. In particular, workers in the plumbing and HVAC sector report higher levels of serious overexertion compared to construction workers in general (Albers et al. 2005). Specific risk factors include the need to drill holes and shoot fasteners to fix overhead hanging systems for piping, HVAC components, and electrical wiring. This work involves using tools that generate high forces, including vibration, rotational, and impact forces. The work requires physical exertion to hold and operate heavy tools, and repetitive activity combined with frequent relocation. Some of this work is performed at ceiling height and can be undertaken in awkward postures, depending on work place size and clearance. These activities create increased risk of WMSDs in the upper extremities (hands, wrists, and elbows), neck, back, shoulders, and knees (Albers et al. 2005). Additional WMSD risks experienced by mechanical and electrical installation workers include the need to unload and transport materials to the location at which they will be used.

8.3.5 Plant Operating

In a review of WMSD risk among operators of construction plant (including cranes, bulldozers, front-end loaders, rollers, backhoes, and graders), Kittusamy and Buchholz (2004) identify whole-body vibration and working in awkward postures as particular risk factors. In an earlier descriptive study by Zimmermann et al. (1997), plant operators most frequently reported WMSD symptoms in the lower back, neck, shoulders, and knees. However, symptoms varied by the type and age of plant being operated, as well as by the length of an individual's work history. Operators with longer work histories, and those working with older items of plant, reported higher incidence of WMSD symptoms. Whole-body vibration is reported to produce systemic effects on the whole body and there is evidence to suggest it causes morphological changes in the lumbar spine (Kittusamy and Buchholz 2004). Kittusamy and Buchholz (2004) report the effects of whole-body vibration are exacerbated when work is performed in awkward postures.

Citing evidence that plant operation involves work in awkward postures (including static sitting), Kittusamy and Buchholz identify the need to quantify and understand the impacts of whole-body vibration and awkward posture in plant operation in construction work settings.

8.3.6 Masonry

Bricklayers and mason tenders are a high-risk group for WMSDs. Bricklayers are frequently required to perform repetitive motions which can result in cumulative trauma disorders or WMSDs. According to Entzel et al. (2007), approximately 60% of injuries caused by overexertion among bricklayers are back injuries. Entzel et al. (2007) also state that major risk factors for back injury among masons involve the weight of bricks, the frequency of tasks, the height at which blocks are picked up and positioned, the height of stands, the distance of blocks from a worker's body, the degree and frequency of twisting involved, and expected production rates.

8.4 Psychosocial Work Stressors and WMSD

There is a growing body of evidence to indicate that WMSDs are linked to psychosocial stressors in the workplace. Psychosocial work stressors have been characterized as:

> … existing circumstances that an individual is exposed to at the workplace and that exert an influence on the individual either through psychologically relevant task organization procedures (e.g. time pressure, job control) or through the social work environment (e.g. lack of social support).
>
> *(Lang et al. 2012, p. 1163)*

The International Labour Organization (ILO) defines psychosocial factors at work as 'interactions between and among work environment, job content, organizational conditions and workers' capacities, needs, culture, personal extra-job considerations that may, through perceptions and experience, influence health, work performance and job satisfaction' (ILO 2016, p. 2). Thus, psychosocial factors reflect the interaction between conditions of employment, the quality and organization of work, and workers' needs, capacities, culture, and family circumstances.

In their systematic review of longitudinal studies investigating the relationship between WMSDs and psychosocial work stressors, Lang et al. (2012) report significant lagged effects whereby the experience of psychosocial work stressors at one point in time significantly predicted the experience of WMSDs at a subsequent time. These significant lagged effects suggest WMSDs are, at least to some extent, caused by exposure to psychosocial work stressors.

Different explanations of the link between WMSDs and psychosocial stressors have been proposed. First, it is believed that psychosocial work characteristics directly influence the biomechanical load experienced while performing a task through change to posture, movement, and exerted forces (Hoogendoorn et al. 2000). It is also believed that psychosocial stressors arising from the organization of work create stress responses which, in turn, impact the musculoskeletal system.

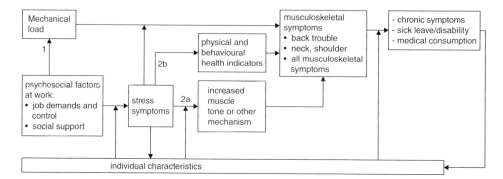

Figure 8.1 Theoretical model linking psychosocial factors at work with stress and WMSDs. *Source:* Bongers et al. (1993).

Several mechanisms have been suggested to explain this relationship. First, stress responses are believed to increase muscle tension which impacts the nervous system, thus compounding the effects of task-related biomechanical strain. Stress responses can also affect hormonal, circulatory, and respiratory functioning which increases physical risk factors. Carayon et al. (1999) suggest stress responses can reduce the body's ability to repair tissue after microtrauma, and can also induce other physiological effects that increase the risk of WMSDs. Eatough et al. (2012) suggest stress responses produce emotional responses (for example, frustration or anger) that can result in risk-taking behaviour (for example, overexertion).

Bongers et al. (1993) developed a theoretical model linking psychosocial risk factors to WMSD. This is reproduced in Figure 8.1.

According to this model there are three potential linking mechanisms. These are:

- *Pathway 1.* Psychosocial factors at work directly influence biomechanical load experienced by workers through changing their posture, movement, and exerted forces – for example, when working under time pressure, movement may be accelerated or undertaken in poor posture.
- *Pathway 2a.* Psychosocial factors at work (that is, demands as well as resources such as job control and social support), combined with individual factors (such as coping ability), can increase work-related stress symptoms which can then increase muscle tone. In the long term this may lead to the development or exacerbation of musculoskeletal symptoms due to some other (unknown) physiological mechanism (for example, hormonal changes).
- *Pathway 2b.* Psychosocial factors at work (that is, demands as well as resources such as job control and social support), combined with individual factors (such as coping ability), can increase work-related stress symptoms. This interaction can then moderate the relationship between biomechanical load and musculoskeletal symptoms due to increasing perceived experience of the symptoms or reducing workers' capacity to cope with them. This could prolong or intensify symptoms of poor health.

Bongers et al. (1993) sought evidence for the linking mechanisms in epidemiological studies. They report inconclusive evidence relating to some psychosocial stressors and explain these are due to the fact that psychosocial and physical workplace risk factors are often closely correlated. Bongers et al. (1993) did, however, find a link between monotonous work, high perceived workload, time pressure, and the experience of musculoskeletal symptoms. They also report low levels of job control and a lack of social support from co-workers are linked with musculoskeletal disease. Further stress symptoms were found to be associated with musculoskeletal symptoms, supporting the position that stress symptoms are a linking mechanism between psychosocial stressors and WMSDs.

This is particularly pertinent in the construction context because the work is physically demanding, and research shows workers experience high levels of stress and burnout (Bowen et al. 2013a,b). In the construction industry, psychosocial factors are shaped by project conditions and pressures embedded in tight delivery schedules, the need to satisfy multiple (sometimes competing) project demands, low levels of control (particularly over work schedules), and the experience of job insecurity. An investigation of the link between psychosocial stressors and WMSDs in the construction industry demonstrates a link between low job satisfaction, high perceived work stress, low job control, and high quantitative demands. Furthermore, worry, distress, and stress reactions not primarily work-related were also linked to WMSD symptoms in construction workers (Sobeih et al. 2006).

High job demands have been consistently linked to WMSDs, particularly when related to working under time pressure (Deeney and O'Sullivan 2009). Time pressure has been identified as a significant factor in the development of chronic low-back pain in construction workers (Latza et al. 2002). Pressure to continuously work, and for this work to be completed urgently, are reported to be particularly relevant predictors of WMSD symptoms in the lower back and upper extremities; that is, the neck, shoulders, arms, and hands (Huang et al. 2003). Further, Huang et al. (2003) report time pressure predicts WMSD symptoms, even after controlling for the presence of biomechanical risk factors. This suggests targeting work time pressures could help to reduce the incidence of WMSDs.

Social support has also been linked to WMSD symptoms of the lower back (Hoogendoorn et al. 2000). A lack of social support is reported to be the most important determinant of burnout and health complaints among construction workers (Janssen et al. 2001). Supervisor and/or work team training focusing on organizational justice and fair interpersonal interaction has been advocated as a strategy that could reduce the incidence of WMSDs (Lang et al. 2012). Hoogendoorn et al. (2000) report strong evidence for low job satisfaction as a risk factor for WMSDs.

Job control has been reported to be directly linked to the experience of WMSDs. Hollmann et al. (2001) also provide evidence that job control has a protective buffering effect for musculoskeletal complaints, but only under conditions of low workload. Thus, when someone's workload is low, the relationship between physically demanding work and WMSD symptoms is non-significant. But, under conditions of high workload, physical demanding work is a significant predictor of WMSD symptoms, irrespective of the level of job control a person has. In this study, Hollmann et al. (2001) define job control fairly narrowly as having control over one's work method and timing. Previous research in the construction industry reports low levels of work-time control as being a work stress factor for project-based workers (Lingard et al. 2012a,b,c).

Hoogendoorn et al. (2000) report strong evidence for low job satisfaction as a risk factor for WMSDs.

Job insecurity is also reported to have a substantial effect on WMSD symptoms (Lang et al. 2012). Job insecurity is a significant concern for most project-based construction workers who can face periods of unemployment between projects (Turner and Lingard 2016a). Sobeih et al. (2009) studied the link between psychosocial factors and WMSD incidence in the USA construction industry and reported the strongest predictors of WMSDs were economic factors, in particular concerns about job security, low wages, and trying to achieve job role expectations. These economic factors significantly predicted musculoskeletal symptoms in the shoulder, elbow, hand, finger, and hip.

The biopsychosocial model of musculoskeletal disorder (described by Deeney and O'Sullivan 2009) also suggests physiological work stress responses may not end after work ceases. Stress responses from work can therefore be maintained and exacerbated by experiences outside work that increase the risk of WMSDs. In situations in which individuals are exposed to further workloads, such as childcare or family/household responsibilities, recovery may be further impacted and the risk of WMSDs increased. This is important because workers in the construction industry are reported to experience high levels of work–family conflict, which is related to poor relationship quality and psychological distress (Lingard and Francis 2004). A supportive workplace is very important in mitigating job-related stressors and Grandey et al. (2007) report that the work–family conflict experienced by male manual, non-managerial workers who work long hours is lower when workers perceive they are working in an organizational environment that supports work–family balance.

Sobeih et al. (2009) recommend dividing psychosocial factors into stressors (risk factors) and moderators (protective factors). Thus, high job demands (for example, long hours) may be mitigated by protective factors in the work environment (for example, a family-supportive work culture). Understanding the interaction between psychosocial factors experienced as being stressful, and those that can help to mitigate the experience of stress, can be helpful in designing strategies to target health and wellbeing and also WMSDs.

8.5 Cultural Influences

In Chapter 5, leadership was presented as a key driver for organizational or workplace cultures that influence WHS. Safety-specific transformational leadership refers to:

- leaders taking an active and inspirational approach to safety issues;
- leaders serving as good models of safety behaviour; and
- leaders encouraging others to work in a safe manner (Kelloway et al. 2006).

Barling et al. (2002) demonstrate that safety-specific transformational leadership is predictive of injuries in work settings. That is, the stronger a supervisor's transformational safety-specific leadership, the lower the reported frequency of occupational injuries.

Eatough et al. (2012) have linked safety-specific leadership to WMSD symptoms in the lower back, shoulder, and wrist/hand. In relation to the lower back and the shoulder, the relationship was fully mediated by psychological strain experienced by workers as a result of their work. This means that when levels of safety-specific leadership are high, workers experience lower levels of psychological strain. This, in turn, reduces the incidence of WMSD symptoms in the lower back and shoulder. Workers whose supervisors are low in safety-specific transformational leadership are less likely to feel supported in their workplaces and more likely to feel pressured to deviate from WHS-related policies and procedures. (This phenomenon was discussed in Chapter 7.)

These results suggest managerial and supervisory safety leadership training may help mitigate the risk of WMSDs in construction and other industries. Leadership development programmes are a recommended strategy for improving workers' physical and psychological health (Kelloway and Barling 2010). In particular, leadership development interventions that promote transformational leadership are likely to be particularly effective as strong and consistent links are reported between transformational leadership style and WHS outcomes (Barling et al. 2002; Zacharatos et al. 2005; Mullen and Kelloway 2009; Smith et al. 2016; Wu et al. 2016). Kelloway et al. (2006) also report negative impacts on WHS performance when leadership is passive or laissez-faire – for example, failing to intervene until problems become serious enough to require attention, or delaying decision making.

8.6 Ergonomic Interventions in Construction

Different types of ergonomic intervention have been used to reduce the risk of WMSDs in the construction industry. Table 8.2 lists some of these.

As shown in Table 8.2, the types of ergonomic interventions deployed in construction vary from technology-based solutions (such as equipment redesign) to administrative approaches (such as worker training and exercise programmes). Rinder et al. (2008) classify ergonomic interventions using two dimensions:

- the extent to which they present short-term or long-term solutions to a problem; and
- the extent to which they are simple or complex to implement.

Figure 8.2 illustrates these dimensions, and the position of different types of intervention in this two-dimensional schema.

Lower left-hand quadrant. Interventions in the lower left-hand quadrant are relatively short-term and simple solutions. They include using protective equipment and small, simple local work area adjustments. They can be easily implemented and do not require substantial organizational investment (Rinder et al. 2008).

Upper left-hand quadrant. These interventions are still relatively simple but implementing them requires longer term planning and investment. These include substitution of hand tools, and ordering materials in different quantities or sizes of packaging to reduce biomechanical impacts.

Lower right-hand quadrant. Interventions in the lower right-hand quadrant may be implemented in the short term, but are more complex to implement, requiring redesign of the way work is organized. These include planning site layout, materials storage, and work sequencing to reduce the need for manual materials handling.

Upper right-hand quadrant. These solutions are referred to as 'evolutionary' – they are longer term and more complex. They reflect the need to engage multiple stakeholders in collaborative processes intended to reduce the risk of WMSDs – stakeholders include architects, engineers and other designers, and manufacturers and suppliers of structures, plant, equipment, and materials. This may require significant effort and

Table 8.2 Ergonomic interventions implemented in the construction context.

Author	Activity	Intervention
van der Molen et al. (2005a,b,c)	• Bricklaying – vertical movement of materials • Bricklaying – picking up bricks and mortar	• Use mechanical lifting devices (crane) • Use trestles or bricklaying scaffold to adjust work height
de Jong and Vink (2002)	• Movement of switch panel cupboards during electrical installation work	• Design and use a mechanical transporting device
Luijsterburg et al. (2005)	• Bricklaying – picking up bricks and mortar, and laying bricks	• Use stools on scaffold to raise bricks and mortar 50 cm above floor level • Use scaffold with split floors to vary height between levels • Use a height-adjustable (hoist-console) scaffold that can be raised according to height of wall being built
Hess et al. (2004)	• Concreting – horizontal movement of concrete over steel reinforcement cage	• Use a skid plate to improve ease of movement of the concrete hose
Ludewig and Borstad (2003)	• General risk of shoulder injury due to overhead work	• Implement an eight-week home-based stretching and strengthening exercise programme
Vink et al. (1997)	• Manual horizontal movement of scaffold components during erection/dismantling • Manual vertical movement of scaffold components	• Reduce size/length of ladders and boards • Develop a pallet truck for movement • Develop an unloading plan • Set out materials in correct order (close to the spot where the scaffold will be built) • Shoulder protection in clothing • Use an electrical winch to raise materials
De Jong and Vink (2000)	• Unloading of glass panels from truck • Horizontal movement of glass panels • Vertical movement of glass panels • Removal of putty	• Design and develop a truck-mounted hoist • Design and develop a glass cart and glass sledge for horizontal movement • Design and develop special scaffolding with a lifting device • Use a truck-mounted crane • Design and develop a mechanical device for putty removal
Burdorf et al. (2007)	• Manual laying of brick-paved road	• Use machine with a hydraulic clamp to pick up and place multiple brick pavers at one time • Use machinery with a vacuum-lift mechanism to pick up and place multiple brick pavers at one time

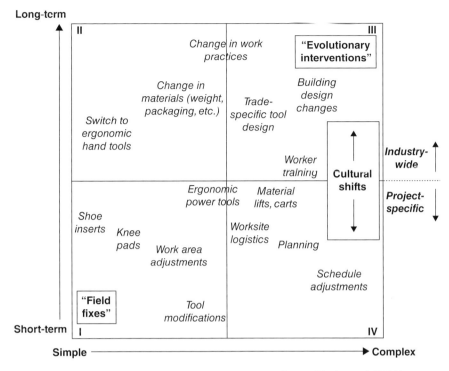

Figure 8.2 Construction ergonomics intervention matrix. *Source:* Rinder et al. (2008).

investment, and may prove challenging given the construction industry's highly frag-mented supply network.

Case Example 8.1 is an example of an evolutionary development in equipment used in the construction industry for overhead drilling.

Case Example 8.1 Ergonomic Intervention for Overhead Drilling

Sustained exposure to overhead construction tasks is strongly associated with pain and musculoskeletal disorders of shoulder, neck, arms, wrist, and back. Overhead drilling into concrete and metal is prevalent in commercial construction. It is a particularly physically demanding task due to the awkward posture, and the strong and continuous upward force required for holding a vibrating hammer drill and pushing it upward for penetration. Other hazards involved in performing this task are exposure to noise and dust, and falls from height (Rempel et al. 2009). To overcome these issues, Rempel et al. (2009) developed and field-tested two intervention devices: an inverted drill press, and a foot lever press. Experienced construction workers were asked to use the devices and compare them with the usual method of drilling used on construction sites. Using a questionnaire, the workers were then asked to assess and rate each method in terms of ease of

---

(Continued)

Case Example 8.1 (Continued)

---

setting up and use, fatigue, and positive and negative features of each device. They were also asked to provide suggestions for design improvement. Fatigue levels were assessed for five body regions: neck, shoulders, hands and forearms, low back, and legs. The study results suggested that, while the intervention devices were believed to cause less fatigue in the five body regions, they were perceived to be difficult to use and to reduce productivity, particularly when moving the device and setting it up. Comparing the two devices, the workers preferred the inverted drill press to the foot press.

In their follow up study, Rempel et al. (2010a,b) developed and tested a second generation of the inverted drill press. To respond to device movement and setup issues, the new design had three different wheeled bases. Similar to the previous study, construction workers were asked to use each device and a questionnaire was used to assess ease of use, fatigue (in five body regions), usability, and positive and negative features of each device. In addition, objective measures were used for productivity, arm loads, and postures. These included monitoring participants, and measuring head extension and hand forces. The field test indicated that participant construction workers preferred two of the designs: a collar base and an adjustable castor base design. Moreover, the design changes were perceived to have improved the usability of the inverted drill, especially in movability and set up time. There was a significant improvement in perceived fatigue ratings when the devices were compared to the usual drilling method. Rempel et al. (2010a) believe the improvement was due to reduced hand forces, reduced shoulder abduction and flexion, and reduced drilling time measured during field tests. During field tests, the feedback from experienced construction workers was vital to improving the intervention devices. Testing devices in field settings, and with construction workers, should be considered when developing new interventions that aim to address occupational risk factors related to construction activities.

Rempel and Barr (2015) have subsequently used feedback from contractors and workers to develop a universal rig for supporting large hammer drills. This was field tested by labourers and electricians who performed their usual work using the new drilling rig and compared this to their usual method of working. Using the new rig, subjective regional fatigue was significant lower in the neck, shoulders, hands, and arms. In terms of stability, control, drilling accuracy, and vibration, workers rated the rig easier to use than their usual manual method of working. Rempel and Barr (2015) also report that using the rig reduced drilling time by approximately 50%.

8.7 Designing for Reduced WMSDs

In the construction industry, the notion of designing for construction workers' safety is now well established (as discussed in Chapter 2). However, the practice of actively reducing risks to health through design is less well established. Designers specify building materials and can play an important role in reducing WMSD risk by considering the weight and mass of materials they specify. For example, Latza et al.'s (2002) study found that laying large lime sandstones was a significant risk factor for chronic low-back pain among bricklayers. Drawing on data collected in the Hamburg Construction Worker Study, Latza et al. (2002) report that bricklaying involves repetitive work, with most of the activity (94.2%) undertaken in a standing position. More than 50% of bricklayers' work hours were spent in a standing and bent position and on average they moved about 881 kg in an hour. Large lime sandstones weighed about 7–10 kg each, and a 40-year-old construction worker laying these sandstones has the same risk of chronic back pain after three years as an unexposed 62-year-old construction worker (Latza et al. 2002). This finding highlights the impact designers' choices can have on workers' musculoskeletal health.

Hess et al. (2010) similarly evaluated the biomechanical impacts of two types of material used in blockwork. They compared the impacts of laying a more traditional concrete masonry unit (CMU) with laying autoclaved, aerated concrete (AAC) blocks. CMU blocks can weigh 16.3 kg and it is estimated a mason lifts 200 CMU blocks in an eight hour work day, equating to over 3260 kg over that time span (Hess et al. 2010). Following field-based measurement of the biometric impacts of laying AAC and CMU blocks, Hess et al. (2010) report significant differences. The AAC blocks were larger and solid, requiring they be lifted with two hands. Although AAC blocks were larger, they were handled for significantly less time because they are laid using an adhesive that is applied not to the block but to the upper edge of the wall being constructed. CMU blocks are held on one side while mortar is applied to them before laying. Hess et al. (2010) note that, although low-back compression forces were higher for AAC than CMU, both exceeded the 3.4 kN recommended upper limit established in National Institute for Occupational Safety and Health (NIOSH) guidelines. In addition, the risk associated with AAC blocks may be mitigated partially by the fact that they are handled for significantly less time, and are larger, so workers would be likely to handle fewer blocks for the same area of wall (Hess et al. 2010). Blocks are selected in different sizes and finishes to meet architectural and structural requirements. Designers play a key role in these selection decisions and can therefore play an important role in influencing the risk of WMSDs in certain high-risk construction trades, such as masonry. Smallwood (2012) recommends improving designers' knowledge through education to enable them to understand the impact of their decisions on WMSD risk and make more informed selection decisions about building materials.

As an example, design professionals responsible for specifying building materials have required greater use of rebar reinforcements for concrete block construction (Inyang et al. 2012). While this allowed them to comply with more stringent building codes, it also led to increased physical workload and affected workers' level of exposure to WMSD risk (Inyang et al. 2012). Thus, it is important to understand that decisions taken to satisfy one project performance requirement can inadvertently introduce WMSD risks.

The design of work also includes the design of an integrated materials management programme that specifies requirements relating to delivery, storage, traffic flow, and mechanized lifting and moving activities. Gervais (2003) argues this will only be effective if materials handling requirements are identified and suitable injury prevention measures are decided upon before construction work commences. For example, Kim et al. (2011) report on the development of a decision support system that incorporates consideration of packaging, transport, delivery sequencing, stacking, and erection sequencing in the design of prefabricated housing components. This system minimized the need to move materials during delivery and onsite construction, and represents an effective method for incorporating WMSD risk reduction into decision making during the design stage of a construction project (Kim et al. 2011).

8.8 Measuring the Risk of WMSDs

There is a growing body of evidence showing the potential for ergonomic interventions to reduce WMSD risk in construction. However, the quality of intervention studies has been criticized (Rinder et al. 2008), particularly methods of measuring performance outcomes. These methods have used back pain monitoring (Holmström and Ahlborg 2005), observational methods (Luijsterburg et al. 2005; van der Molen et al. 2004), videorecording and three-dimensional strength prediction (Vink et al. 2002), the NIOSH Lifting Equation to estimate the risk inherent in a manual handling task (Mirka et al. 2003), and self-report survey methods (Holmström and Ahlborg 2005; Ludewig and Borstad 2003; Vink et al. 2002). Rinder et al. (2008) argue these methods provide information linking job task characteristics (such as flexion, rotation, bending, velocity, acceleration, and compression force) with intermediate measures of musculoskeletal health. However, they do not provide objective measures of the physical risk factors inherent in the work. Consequently, Rinder et al. (2008) recommend using biomechanical assessments to better understand the implications of task design on human body dynamics and the risk of WMSDs.

The use of wearable sensors to capture biomechanical data in field settings has become feasible and relatively cost-effective. Hess et al. (2004) used a custom-designed sensor system to measure workers' lumbar region posture, motion, and force during concreting operations. More recently, proprietary systems capable of field-based measurement of full-body movement have become available and are being deployed in ergonomic assessment and evaluation projects in the construction industry (see, for example, Brandt et al. 2015). In Chapter 6, indirect measures of WHS, including incidence rates of injury and illness, were criticized because they are after-the-fact measures of negative events that may or may not be predictive of future performance (Dekker 2014). As such, their validity and usefulness are limited when it comes to informing evidence-based strategies for WHS improvement. Objective and direct measurement of ergonomic risk factors can be used to properly evaluate risk reduction strategies for WMSDs and provide an important body of evidence underpinning ergonomic solutions in the construction industry. Case Example 8.2 provides an example of site-based measurement of the risk of WMSDs in the construction industry.

Case Example 8.2 Measuring the Risk of WMSDs in Steelfixing

Placing and securing steel bars used in reinforced concrete involves heavy manual materials handling and work in awkward postures. International research shows steelfixers spend 40% of their work time in awkward trunk postures (Buchholz et al. 2003). Steelfixing also involves a high risk of work-related musculoskeletal injury affecting the hand, wrist, or fingers, with up to 48% of steelfixers reporting symptoms in these areas of the body (Forde et al. 2005).

A team of researchers from RMIT used a whole-body system of wearable sensors to capture information about risk factors for WMSDs in steelfixing. Researchers visited rail construction sites within the Victorian Government's Major Transport Infrastructure Program.

Case Example 8.2 (Continued)

Construction workers participated in the study by wearing a whole-body system of lightweight sensors to objectively measure movement of the muscles and joints while workers were performing their daily work tasks. Workers also provided feedback on work methods, tools, and equipment.

Preliminary measurements found that hotspots for musculoskeletal injury in steelfixing are the back, the shoulder, and the wrist.

Some ergonomic tools have been developed to reduce the amount of bending involved in fixing steel below knee heights. The study evaluated the impact of using three different tools for steelfixing. These were: a conventional pincer-cutter tool, a power-tying tool, and a long-handled stapler tool.

The Back

The conventional pincer-cutters and the power-tying tool did not differ in terms of the extent to which steelfixers had to work in a bent posture when fixing steel at lower work heights. However, the long-handled stapler tool significantly reduced bending of the back when working below knee level (Figure 8.3).

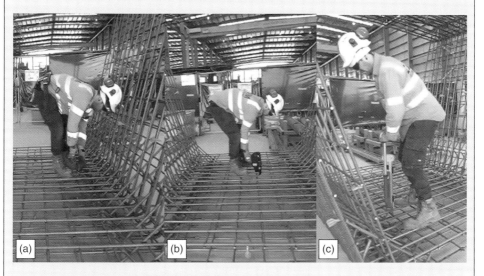

Figure 8.3 Steelfixing at ground level with (a) a conventional pincer-cutter tool, (b) a power-tying tool, and (c) a long-handled stapler tool.

The average trunk (back) inclination was reduced from 74° using conventional pincer-cutters to 34° using the long-handled stapler tool when fixing steel at ground level (Figure 8.4).

(Continued)

Case Example 8.2 (Continued)

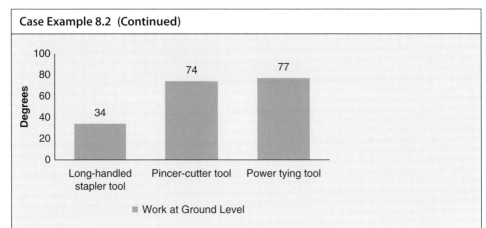

Figure 8.4 Average trunk inclination using different tools for fixing steel at ground level.

The Wrist

Conventional steelfixing involves repeated rotation of a pincer-cutter tool to twist and tighten wire. Repeating this action over a working day increases the risk of wrist injury because it involves repetitive twisting and turning.

When using the power-tying tool, the wrist was almost straight when working at both knee-to-hip and hip-to-shoulder heights (indicated by range of movement values close to zero in Figure 8.5). In contrast, the pincer-cutter tool involved significantly greater bending of the wrist (indicated by high positive values), increasing the risk of WMSDs affecting the wrist (Figure 8.5).

The long-handled stapler tool performed differently in terms of the risk of wrist injury depending on whether steelfixing was undertaken at knee-to-hip or hip-to-shoulder height. When work was at knee-to-hip height, a steelfixers' wrist remained relatively straight when using the long-handled stapler tool. However, when the height of work moved to between the hip-to-shoulder, the long-handled stapler produced a greater range of wrist movement, which could be hazardous.

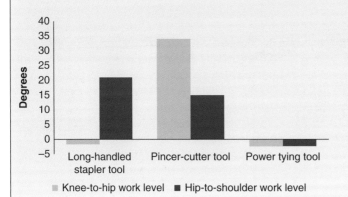

Figure 8.5 Peak right wrist flexion/extension. Positive values indicate wrist flexion and negative values indicate wrist extension.

Case Example 8.2 (Continued)

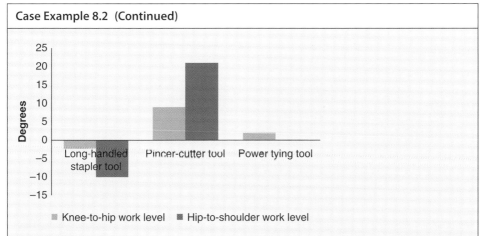

Figure 8.6 Peak right wrist rotation. Positive values indicate wrist pronation and negative values indicate wrist supination.

The power-tying tool also significantly reduced the amount of wrist rotation when fixing steel (Figure 8.6). The values close to zero reflect that, when steel was fixed using the power-tying tool, there was little or no rotation of the wrist.

The Shoulder

Working overhead uses awkward shoulder postures and movements that may lead to shoulder injury. Work overhead should be avoided wherever possible. However, if work overhead cannot be eliminated, the use of a long-handled stapler tool reduces awkward shoulder movements, reducing the risk of shoulder injury (Figure 8.7).

Figure 8.7 Steelfixing above shoulder height with (a) a conventional pincer-cutter tool, (b) a power-tying tool, and (c) a long-handled stapler tool.

(Continued)

Case Example 8.2 (Continued)

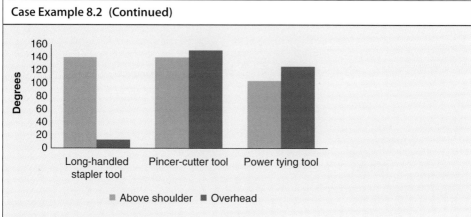

Figure 8.8 Peak right shoulder elevation.

Figure 8.7 shows the high range of shoulder movement involved in performing steel-fixing work above the shoulder and overhead. A high range of movement presents a high level of risk of WMSDs affecting the shoulder. When work is performed overhead, the range of shoulder movement was 151° when using conventional pincer-cutters. This reduced to 13° when the long-handled stapler tool was used, significantly reducing the risk of WMSDs in the shoulder (also shown in Figure 8.8).

Wherever possible, steelfixing tasks should be designed to avoid awkward postures, excessive bending of the back, or work above shoulder height. Consideration should be given to the height at which steel bars are to be fixed and, where possible, work should be designed to reduce bending of the back (for example, when working below knee level) or extension of the shoulders (for example, when working above shoulder height).

Where this is not possible, care should be taken in selecting the most ergonomically effective tools for the job. The research showed that tools specially designed for steelfixing can make a difference in reducing the risk of work-related musculoskeletal injury when fixing steel.

But no single steelfixing tool was ideal in all situations. While the long-handled stapler tool reduced bending of the back while fixing steel at lower work heights, the stapler tool required the use of a forceful pushing and pulling action to fix and twist the wire tie. The power-tying tool significantly reduced the repetitive movement of the hand and wrist, but was heavy to hold in one hand and did not reduce the need to bend the back while working below knee level.[1]

---

1 This research was jointly funded by WorkSafe Victoria and the Major Transport Infrastructure Program, Department of Economic Development, Jobs, Transport and Resources, Victorian Government.

Chapter 6 discussed problems associated with focusing too heavily on things that have already gone wrong in a work system, thereby framing WHS as a negative construct. Chapter 7 also discussed different approaches to human variability in work systems and whether this should be seen as a problem to be controlled, or as a source of creativity and innovation. Vink et al. (2006) argue that, in relation to workplace ergonomics, it is preferable to focus on health rather than illness. Vink et al. (2006) argue that overly prescriptive statements of what is not allowed can be demotivating and stifle innovation in ergonomic interventions. A better approach is to focus attention on what can improve the health of workers, as well as other aspects of overall system performance. This change in focus is also consistent with the following definition of ergonomics provided by the International Ergonomics Association:

> Ergonomics (or human factors) is the scientific discipline concerned with the understanding of interactions among humans and other elements of a system, and the profession that applies theory, principles, data and methods to design in order to optimize human well-being and overall system performance.
>
> *(cited in Vink et al. 2006, p. 537)*

8.9 Participatory Ergonomics

In Chapter 7 we discussed the potential benefits associated with engaging workers in designing work processes and developing WHS-related procedures. That perspective is relevant here because WMSDs have been linked to low levels of participatory management, in particular for people experiencing both lower back and upper extremity symptoms (Huang et al. 2003). This finding may provide an important insight into the potential to use a participatory ergonomics (PE) approach for reducing WMSDs. PE relies on involving workers in developing ergonomic work procedures and provides a framework through which workers and managers can engage in open conversation about work design and organization (Rivilis et al. 2006). In a PE approach, ergonomists act as change agents, rather than technical experts. Hess et al. (2004) describe how their particular application of a PE approach was guided by cooperative inquiry. In contrast to typical biomedical models in which technical specialists determine and control intervention design and implementation, cooperative inquiry elicits and integrates the knowledge of research subjects (for example, workers) into decision making and evaluation.

Prior research has demonstrated that a PE approach can reduce the risk of WMSDs (Rivilis et al. 2008). Vink et al. (1997) describe a PE process they deployed in two scaffold building companies. This process follows six steps, described below:

1. Workers are informed of the PE project and advised that the project's aim is to improve working conditions.
2. An analysis is undertaken of risk factors inherent in work tasks targeted for improvement.
3. The analysis is presented to workers and other relevant stakeholders and ideas for improvement are elicited.
4. Ergonomic measures to be implemented are jointly agreed and subject to a pilot implementation. Adaptations are made if necessary.
5. Ergonomic measures are then introduced more broadly to workplaces and supported by promotion, training, and instruction.

6. Workers and other relevant stakeholders jointly evaluate the impact of the ergonomic measures (Vink et al. 1997).

In their study of scaffolders, Vink et al. (1997) report that workers identified one of the most serious issues they face to be shoulder complaints related to vertical and horizontal movement (by manual handling) of scaffold components. Consequently, a number of measures were introduced to reduce the risks associated with the manual handling of components. These included establishing a maximum ladder length (3 m), developing a pallet truck to transport scaffold members, providing electrical winches for vertical lifting of scaffold components, and setting a maximum length of scaffold boards (4 m). Vink et al. used heart rate measurements, observational methods, and a survey to evaluate the impact of the revised methods. They report:

- a significant reduction in heart rate;
- dramatic decrease in the percentage of time scaffolders were observed lifting or carrying weights of more than 20 kg;
- reduced percentage of time workers were observed working with a shoulder elevation of more than 60° (considered a risk factor for shoulder symptoms); and
- participants considered the PE process to be effective.

PE has also been applied in a number of other studies about construction trades. For example, Hess et al. (2004) used PE to implement and evaluate a skid plate device for the horizontal movement of concreting. Importantly, they observed that using the skid plate initially increased flexion significantly and did not change other movement risk factors related to the risk of lower back disorders. It was only after workers modified the skid plate by attaching it to the concrete hose that risk factors for lower back disorder were reduced. The overall probability of lower back disorder was estimated to have reduced from 67% prior to skid plate use to 46% when using the worker-modified, secured skid plate (Hess et al. 2004). This finding highlights the role played by local adaptation or 'field fixes' to ensure ergonomics interventions have their desired effect.

Integrating construction workers' insights and experiences into designing, implementing, and evaluating ergonomics interventions can produce more effective interventions and a valid evaluation of these interventions' practical impact (Williams et al. 2010). Further, involving workers in PE processes helps to build their knowledge and awareness of WMSD risks and increase the likelihood that any changes or improvements will be adopted and sustained (Moir and Buchholz 1996). However, PE advocates argue that, to be effective, a PE process requires strong management commitment and support, and genuine worker engagement and involvement (Vink et al. 1997).

Vink et al. (2006) suggest a PE process can involve:

- consultation with workers (that is, workers are informed about the process but decisions are ultimately made by ergonomists or managers); or
- the empowerment of workers (that is, workers make decisions and have control over the changes made to their workplace, equipment, or work process).

The impact of empowerment on the success of PE approaches was explored by Vink et al. (2006) who found that, in some cases, empowerment did not produce success. This was particularly true in complex work situations in which workers (and even their employers) did not have enough influence to change a work situation. One such circumstance was observed in construction in which the equipment provided by the principal contractor restricted a subcontractor's ability to implement ergonomic interventions (van der Molen et al. 2005a,b,c). This illustrates the need to engage participants at multiple levels in the hierarchy of contractors, subcontractors, and suppliers in order to implement PE approaches effectively in the construction context. Vink et al. (2006) did find some evidence to suggest PE interventions are more likely to succeed when workers are actively engaged in the redesign process, are able to experience the different ways of working, and feel they have some control over changes that are made

8.10 Conclusions

This chapter describes the complex aetiology of WMSDs. It is evident that multiple construction project stakeholders can play a part in reducing WMSD risks. Psychosocial risk factors for WMSDs include working under time pressure. Construction work is very schedule driven – and there are severe financial penalties for failing to complete work on time. Clients play a key role in determining project timelines and principal contractors respond by mirroring client contractual requirements in subcontract agreements. Thus, work schedule pressures are pushed down the contracting chain, and are ultimately borne by workers who undertake the work. The impacts on WMSD risk of schedule demands, and other psychosocial work stressors, should be carefully considered when developing project delivery schedules and establishing resourcing arrangements. Broader aspects of the quality of employment and jobs should also be considered, including job security, levels of control and/or support available to construction workers, and the prevailing culture of the industry. Focusing on improving the quality of work could reduce stress responses that have been linked to WMSDs in various ways. Attributes of project and organizational cultures also linked to WMSDs include the extent to which leadership is supportive of workers' health, safety, and work–family balance. Targeting these aspects of workplace culture should form part of a holistic approach to creating an environment that enables (rather than impedes) optimization of wellbeing and performance.

Considering WMSD risk reduction in design is also recommended. Design professionals make choices about building structural elements, building technologies, materials, and finishes; these choices can all impact on the risk of WMSDs. For example, careful consideration of the physical properties (size, mass, and weight) of specified components and materials could help to reduce risks associated with manual materials handling. However, there may be many other ways designers could creatively contribute to reducing WMSD risk; for example, selecting building systems that reduce the need for physically demanding manual processes, and/or the use of equipment that exposes workers to high levels of hand/arm or whole-body vibration.

Through paying attention to ergonomic design, designers, manufacturers, and suppliers of construction tools, plant, and materials can also play a part in reducing WMSDs. To ensure workers' tacit knowledge about locally situated practice is incorporated into decision making, a participatory ergonomics (PE) approach is highly recommended. Engaging workers in developing, implementing, and evaluating WMSD risk reduction interventions is likely to produce sustainable improvements that are practical and accepted by workers. Research evidence suggests local adaptations of ergonomic interventions can improve their performance and adoption.

Thus, reducing WMSDs in the construction industry requires an interdisciplinary and collaborative approach: clients, designers, constructors, suppliers of plant, equipment, and materials, and workers all need to collaborate to develop, evaluate, refine, and implement solutions. This will not be easily achieved given the fragmented construction industry supply chain. However, in the context of an ageing workforce and strong evidence that WMSDs are a leading cause of work disability and early retirement among construction workers, the industry has much to gain by responding to the frequency of WMSDs in ways that are more strategic, integrated, evidence-informed, and effective.

Discussion and Review Questions

1   Why are work-related musculoskeletal disorders (WMSDs) such a significant and persistent WHS problem in the construction industry?

2   What opportunities are there to consider the design and quality of work in the construction industry to reduce the incidence of WMSDs?

3   How can different industry participants contribute to reducing WMSDs in construction projects?

4   What opportunities are offered by a participatory ergonomics approach to reducing WMSD, and how can this be implemented in construction projects?

Acknowledgements

The research presented in this chapter was jointly funded by WorkSafe Victoria and the Major Transport Infrastructure Program, Department of Economic Development, Jobs, Transport, and Resources, Victorian Government.

Sam Rahnama's research assistance in developing a literature review is also acknowledged.

9

Considerations for the Future of Construction Work Health and Safety

9.1 Rethinking Traditional Ways of Working

Many hazards are present in the construction work environment. Worksites are constantly changing, with continuous movements of mobile plant and equipment, materials and people. In vertical building work, as well as in the construction of elevated bridge structures, work is performed at height, with the risk that people could fall or objects could be dropped to lower levels. Physically demanding manual work tasks are still required in many construction processes, increasing the risk of work-related musculoskeletal injury. Construction workers are also exposed to a vast array of chemicals and to potentially harmful respirable dusts, noise, and vibration. As was noted in Chapter 4, in comparison to safety hazards, occupational health hazards are less effectively managed in the construction industry. Yet, the relative neglect of occupational health risks is alarming due to the sheer number of people seriously affected. In the UK, for example, it is estimated that construction workers are at least 100 times more likely to die from a disease caused or made worse by their work as they are from a work-related injury (IOSH 2015). Although construction industry organizations are beginning to focus more effort on preventing occupational illnesses (see Hopkinson et al. 2015, for an example), significant barriers still exist to the effective control of occupational health hazards at source, or through identifying risk mitigation measures at the point of design decision making. This is illustrated in Case Example 9.1.

Case Example 9.1 A Foundation System Example of Reducing Occupational Health Hazards

Traditionally, breaking down the tops of concrete piles to expose steel reinforcement bars has been carried out using hand-held pneumatic breakers (see Figure 9.1).

This method of pile breaking involves several serious occupational health hazards, including exposure to hand arm vibration, dust, noise, and the risk of work-related musculoskeletal disorders (WMSDs).

A recent study in the Australian construction industry revealed that a hand-held pneumatic breaker is still routinely used for breaking down pile heads. Further, the specified method of risk control was not reliable. A layer of non-bonding material (foam) was to be

---

Integrating Work Health and Safety into Construction Project Management, First Edition.
Helen Lingard and Ron Wakefield.
© 2019 John Wiley & Sons Ltd. Published 2019 by John Wiley & Sons Ltd.

installed at the desired 'cut-off point' during construction of the concrete piles. This material had to be installed before the concrete was poured. If installed correctly, incorporating the non-bonding material significantly reduces the length of time it takes for the pile to be broken mechanically with a hand-held pneumatic breaker. However, onsite observation revealed that pile construction work and pile breaking work were undertaken by two different subcontracted work crews. In many instances, the construction crew did not install the non-bonding material correctly, resulting in a substantial increase in health risk exposure for the workers who subsequently need to break down the concrete pile heads.

Figure 9.1 Using a hand-held pneumatic breaker to break down concrete pile heads.

However, if considered at the design stage of a project, technological risk controls are available. Commercially available hydraulic pile-breaking technologies are available to eliminate the need for breaking using hand-held pneumatic tools (EFFC 2015). In addition, 'integrated' active systems have been developed. These systems incorporate an active pile-breaking system within the pile. This system is activated when the concrete is cured. An example of an integrated active pile-breaking method is the 'Recepieux' system, which:

- uses a system of breakers installed at the desired cut-off position before pouring concrete (Figure 9.2a);
- introduces an expanding agent into the pile (at least 72 hours after pouring) through carefully positioned ducts which deliver chemicals to the breakers (Figure 9.2b), enabling the pile top to be cut off within the ground, the breakers working like a jack (Figure 9.2c); and finally
- enabling the pile top to be mechanically lifted off without the need for jackhammering (Figure 9.2d).

This case example also highlights the importance of considering the work health and safety (WHS) aspects of alternative work processes (such as pile breaking methods) when designing features of the permanent structure (in this case a foundation system). Ensuring detailed knowledge about construction processes (and their WHS implications) is available to product designers was flagged in Chapter 3 as a critical success factor for safety in design.

But perhaps the most telling thing about Case Example 9.1 is that the health risks associated with mechanical breaking of concrete piles using a hand-held pneumatic tool have been known for more than a decade. Previously, alternative ways to

Figure 9.2 The Recepieux pile-breaking method. *Source:* images reproduced with permission.

undertake this work process, significantly reducing the risk of occupational injury and illness, were described by Gibb et al. (2007). It is the fact that these alternative ways have not replaced mechanical breaking methods that is noteworthy.

The construction industry's strong cultural adherence to traditional ways of working – even though these sometimes have serious consequences for workers' health or safety – is also reflected by low levels of innovation and the industry's slow adoption of new tools, technologies, materials, and work methods (van der Molen et al. 2005a). For example, Kramer et al. (2010) considered technologies, tools, materials, and processes with the potential to improve WHS in the construction industry and identify industry-level and organization-level factors impeding the adoption of new ways of working. These factors include a low level of awareness of risks (particularly those relating to occupational health), a lack of knowledge about alternative ways of working that could reduce WHS risk, and a reliance on informal communication networks for spreading knowledge. Dubois and Gadde (2002) explain the construction industry's slowness to adopt change in terms of the combination of loose and tight couplings in the industry's supply arrangements. Thus, in temporary project supply networks, interdependence and uncertainty promote creativity and localized problem solving. However, relatively loose coupling between firms and projects impedes corporate learning from project experiences. Further, the industry's permanent supply network is characterized by short-term, market-driven transactions between firms. This characteristic does not encourage inter-firm cooperation or implementation of partner-specific adjustments to materials, methods, products, or processes.

9.2 Dealing with Emerging Issues

In addition to long-recognized and persistent problems, the construction industry is increasingly affected by emerging issues associated with the changing nature of work and demographic trends. These issues require that a broader view of the factors that impact workers' WHS is taken.

The construction industry has long relied heavily on subcontracting, which has been linked to poor WHS outcomes (Manu et al. 2013). Loosemore and Andonakis (2007) argue that, although trade subcontractors make up the bulk of the Australian construction industry's workforce and often account for over 90% of a project's value, they 'lack the resources, culture and skills' to manage workplace safety risks effectively (p. 580). Wadick (2010) suggests poor communication between trades, and ineffective consultation between workers and managers, increase safety risks associated with subcontracting in construction projects. Further, 'payment-by-results' arrangements under which subcontractors are typically engaged can encourage corner-cutting with regard to workplace safety (Mayhew and Quinlan 1997; Mayhew et al. 1997). Increasingly the construction industry is also using labour hire agencies to meet its workforce needs. Since the 1990s, the number of temporary workers employed by labour hire agencies and 'placed' at host or client worksites has increased dramatically across the member states of the European Union, as well as in Australia and the USA (Underhill and Quinlan 2011).

Underhill and Quinlan (2011) undertook a study of temporary agency workers in Australia. They reported that, compared to directly employed workers, temporary agency workers were more likely to have insecure employment, contingent wages, and long or irregular hours. These differences contributed to agency workers working while injured, potentially exacerbating their injuries, often until they were physically unable to continue working. Agency workers also reported higher levels of work intensification and increased exposure to risk as a result of staff shortages and cost cutting.

Agency workers were also more likely to be inexperienced, young, and unfamiliar with tasks they were required to complete. In some cases, workers who were physically unfit for demanding manual work were allocated to labouring jobs, while others did not have specialized skills required to perform tasks safely. Both these types of mismatch contributed to injury. The agency workers also received less training than directly employed workers, were less involved in workplace consultation and received less communication about WHS. Agency workers had limited protection from arbitrary dismissal and were reluctant to 'voice' WHS concerns or seek assistance from trade union representatives (Underhill and Quinlan 2011).

LaMontagne et al. (2012) report that workers engaged in precarious work experience work-related psychosocial risk factors to a greater extent than workers engaged in more secure forms of employment. 'Work-related psychosocial risk factors' is a term used to describe the 'social and relational aspects of work design that have the potential to produce detrimental effects on employee psychological (e.g. stress, burnout, depression) and physical health (e.g. musculoskeletal disorders, cardiovascular disease)' (Potter et al. 2017, p. 91). Work conditions that can produce psychosocial risks are:

- excessive workloads;
- conflicting demands and lack of role clarity;

- lack of involvement in making decisions that affect workers;
- lack of influence over the way a job is done;
- poorly managed organizational change;
- job insecurity;
- ineffective communication;
- lack of support from management or colleagues;
- psychological and sexual harassment; and
- third-party violence.

Psychosocial risk factors and work-related stress are a significant occupational health issue. They are a key component of the European Union's Survey of Enterprises on New and Emerging Risks (EU-OSHA 2012). The presence of work-related psychosocial risk factors in the construction industry has been recognized for some time and could therefore be said to have well and truly emerged as an occupational health and safety phenomenon. For example, a six-year cohort study of bridge and tunnel construction workers who worked round the clock, long hours, and long weeks, had mortality comparable to other construction workers but were treated more often in hospitals for infectious and parasitic diseases, diseases of the nervous system, diseases of the circulatory system, diseases of the respiratory system, diseases of the digestive system, and diseases of the musculoskeletal system and connective tissue (Tüchsen et al. 2005). Construction workers' concerns about job insecurity have also been linked to poor self-reported levels of mental and physical health (Turner and Lingard 2016a). Links between work-related psychosocial risk factors and health outcomes in construction workers are already well established, with evidence suggesting associations with lower back pain (Holmström et al. 1992; Latza et al. 2002), mental health complaints (Boschman et al. 2013), injury and/or near-injury experiences (Abbe et al. 2011; Goldenhar et al. 2003), and WMSDs (Engholm and Holmström 2005).

The construction industry has a long way to go in developing and implementing management approaches to tackle psychosocial risk factors. However, construction is not alone in this respect. Leka et al. (2015) argue that work-related psychosocial risk factors are poorly understood and management strategies implemented for them often focus on 'mending harm' rather than preventing it from occurring in the first place. Further, because work-related psychosocial risk factors are inextricably linked with the way workplaces are managed and with power relations within workplaces, there is some resistance to addressing these risk factors within workplaces as this is perceived to interfere with managerial prerogative (Jespersen and Hasle 2017). Notwithstanding this, managerial decision making drives the way work is done in construction organizations and projects and can have adverse health impacts for the workforce.

9.3 Improving the Quality of Construction Jobs

The quality of a job reflects both the quality of employment and quality of work associated with that job. Employment quality refers to those aspects of the employment relationship that have a potential impact on the wellbeing of workers: these are all the aspects related to the employment contract – remuneration and working hours, and career development. Work quality refers to how the activity of work itself, and the

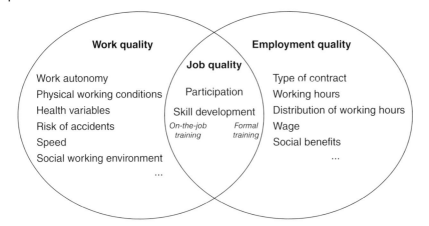

Figure 9.3 A general model of job quality. *Source:* adapted from Bustillo et al. 2009, p. 14.

conditions under which it takes place, can affect the wellbeing of workers: this includes autonomy, intensity, social environment, and the physical environment.

Exposure to the risk of accidents is generally posited as a component of work quality (see Figure 9.3).

However, it is increasingly recognized that many other aspects of job quality are significantly linked both to health and to workplace safety outcomes.

For example, the Job Demands-Resources (JD-R) model has been used to explain links between (i) job demands and resources and (ii) workplace safety (Nahrgang et al. 2011). According to the JD-R model, working conditions include particular job demands and also resources (Bakker and Demerouti 2007; Demerouti et al. 2001). Job demands are 'those physical, psychological, social, or organizational aspects of the job that require sustained physical and /or psychological (cognitive and emotional) effort or skills' (Bakker and Demerouti 2007, p. 312). Job demands include such things as poor work relations, physically demanding tasks, or long work hours, and result in physiological and/or psychological costs to workers who experience them.

On the other hand, job resources are 'those physical, psychological, social, or organizational aspects of the job that are either/or: functional in achieving work goals; reduce job demands and the associated physiological and psychological costs; stimulate personal growth, learning, and development' (Bakker and Demerouti 2007, p. 312). Job resources could include income security, social support from supervisors or peers, and the ability to actively participate in decisions made about how work is done.

Nahrgang et al. (2011) examined the extent to which job demands and resources affected occupational safety outcomes (the incidence of adverse events, accidents, and injuries). They report significant relationships and two mechanisms at play. First, a health impairment process is evident through which workers' exposure to high job demands contributes to impaired health and burnout, which in turn, contribute to the experience of accidents and injuries. Second, job resources increased workers' engagement in safety-related behaviour and also mitigated their experience of health impairment and burnout. These results suggest providing working conditions characterized by manageable demands and sufficient resources to help workers meet these demands, is likely to have positive benefits for workers' health, safety, and engagement with WHS.

In a similar analysis, Hansez and Chmiel (2010) also found the JD-R model explained workers' safety behaviours. Job demands predicted job strain which, in turn, predicted the occurrence of routine safety rule violations. In contrast, job resources were linked to higher levels of worker engagement and fewer rule violations. Another interesting aspect of Hansez and Chmiel's findings was that managers' commitment to WHS could be distinguished from job-related effects on safety behaviour. Thus, the quality of work, in terms of the presence of job demands and availability of resources, operates to shape safety-related behaviours independently of management activity specifically focused on WHS.

In construction organizations and projects, the nature of work demands, and the availability of resources, are clearly shaped by managerial decisions about such activities as procurement, employment, tendering, project resourcing, planning and scheduling, supervision, and workforce development. Improving the quality of construction jobs is likely to both produce benefits in terms of WHS performance and help to attract skilled workers to construction. Potentially, it could also increase participation of currently under-represented groups of workers within the industry.

9.4 Managing WHS as an Integral Part of Work

Since the 1990s there has been a growing emphasis on systematic management of WHS. Frick and Wren (2000) draw a distinction between traditional approaches to WHS and WHS management. WHS management is deemed to differ from traditional, reactive WHS in 'emphasising from the outset the quality control of WHS through managerial acceptance of responsibility, integration and systematic management of production' (p. 19) with the objective of identifying injury and illness hazards early in production processes and introducing preventive strategies and controls before injury or illness occur.

Frick and Wren also differentiate systematic management of WHS from WHS management systems. They contend that WHS management systems establish rules about how to manage WHS that extend far beyond the principles of systematic management. WHS management systems (sometimes referred to as safety management systems, reflecting the tendency to downplay occupational health as an area of organizational performance in need of systematic management) are not always clearly or consistently defined. However, these systems typically emphasize process control, planning, documentation, system performance monitoring, and feedback (Li and Guldenmund 2018). Li and Guldenmund also suggest WHS management systems effectively 'bundle' all WHS-focused management activities in 'an orderly manner' (p. 96).

One potential consequence of this bundling is that the management of WHS can become disassociated from the operational management of construction work. The professionalization and segregation of WHS management can, in turn, substantially impact the extent to which site-based workers, supervisors, and others develop a sense of responsibility for, and 'ownership' of, WHS (Sherratt 2016).

In particular, pressure to make WHS visible and auditable has encouraged the prevalence of generic 'tick and flick' management systems that do not reflect the way work is practised within workplaces (Hohnen and Hasle 2011). For example, an analysis of WHS in the Australian construction industry identified the purchase of 'off the shelf' generic systems as an impediment to effective local management of WHS (Lingard et al. 2017b).

WHS management systems, including those commercially marketed by consultants and private companies, also emphasize achieving standardization and eliminating human error through establishing context-free work procedures, rules, and instructions (Reiman and Rollenhagen 2011). Yet, in reality, all behaviour at work is contextual and responsive to local conditions. People do what makes sense in a given situation, and sometimes this is not what is prescribed in documented WHS procedures. This is particularly the case when these procedures have been developed without significant input from people who perform the work. One of the main criticisms levelled at the operation of WHS management systems is that planning and developing procedures, work instructions, and rules is not always informed by the experiential knowledge of people who perform the work (see Antonsen et al. 2008). It is likely that the gap between 'work as imagined' and 'work as done' will be greater in the decentralized and fragmented construction industry in which subcontractors are required to comply with principal contractors' corporate WHS policies and procedures, and sometimes work under significant commercial pressure.

To ensure WHS management is systematic and effective, the interaction between elements of the WHS management system and other areas of operational management should be considered. When important decisions are being made about many aspects of business and project management, the potential WHS implications of those decisions should be properly thought through and appropriately managed. Seeking input from people who perform construction work, including specialist subcontractors, in developing and reviewing work procedures can also help ensure WHS-related rules make practical sense: it is an approach that will ultimately encourage compliance.

9.5 Focusing Effort Where It Matters

In some circumstances, WHS management systems and organizational WHS programmes can dedicate an unnecessarily large amount of time and effort to things that provide little benefit in improving WHS. Worse still, some elements of WHS management systems, and/or widely implemented organizational safety programmes, may cause more harm than good. Dekker (2014) describes the bureaucratization of WHS that has occurred since the 1970s, involving:

- establishment of hierarchical structures, including reporting structures and accountabilities;
- a specialization and division of labour; and
- constraint and control behaviour through establishment of formalized rules and procedures.

Bureaucratization has brought benefits by improving standardization and WHS performance. However, questions are now being raised about the role and impact of bureaucratic WHS processes and structures, particularly those related to proliferation of WHS-related rules and an obsession with quantifying WHS performance. For example, concerns have been raised about the sheer volume of WHS-related documentation produced within WHS management systems. Some research suggests the amount of WHS paperwork managers are expected to deal with reduces their availability for 'hands-on' supervision of work (Lamvik et al. 2009).

Also, critical information that workers need can be 'buried' inside long, overly complicated documents (Bieder and Bourrier 2013). Weichbrodt (2015) observes that WHS-related rules grow in volume and complexity as accident investigations create public pressure to take preventive action. Yet, when new rules are established, old rules are usually not reviewed, revised, or removed. However, adding new rules does not necessarily improve WHS. Rather, research has shown that in over-proceduralized work environments, workers are insufficiently mindful to respond to emergent dangers and dynamic workplace conditions (Fucks and Dien 2013).

In the construction context, excessive documentation presents particular challenges. In an industry in which time is money, principal contractors do not provide sufficient time for subcontracted workers to read WHS documents before commencing work (Lingard et al. 2015b). Further, low levels of literacy significantly impact some workers' ability to comprehend information provided in written form. Yet, construction workers are routinely expected to 'sign off', stating they have read and understood the content of these documents before commencing work (Lingard et al. 2015b). Effectively this is no more than a risk mitigation strategy for the principal contractors (and an attempt to transfer the liability for any injury or harm onto workers). It cannot claim to produce genuine benefits for workers' health or safety.

In recognizing problems associated with ensuring workers understand important WHS-related rules, some construction companies are streamlining their WHS documentation and supplementing it with visual content, such as images or video.

So-called 'Zero-Target' programmes have also come under criticism for placing too much emphasis on reducing to zero a single lagging safety indicator: the lost-time injury frequency rate (Dekker 2017). This is practically problematic in the context of inherently dangerous construction work. It also has the potential to create more harm. Heavy emphasis on targets and numbers can result in suppressing accident reporting and failure to prevent serious long-term diseases, especially when external motivations (brand image, for example) drive the programme (Frick 2011). In an ethnographic study of UK construction sites, Sherratt (2014) describes how Zero-Target programmes can produce disenchantment and disengagement when targets are not supported with significant changes to the conditions within which work takes place. Further, Zero-Target programmes can be dismissed, quickly and easily, if a single incident occurs.

Dekker challenges the idea that the only way to move towards reducing harm (or even to zero) is to focus managerial effort on preventing incidents and injuries. Drawing on concepts from resilience engineering, he proposes a more effective way to maximize intended outcomes: to understand and actively foster capabilities that ensure things go right in the workplace. In particular, Dekker counsels that a more cogent approach to effective long-term safeguards for system safety is to nurture capabilities that enable people to adapt to adverse or unexpected events in ways that enable a system to operate safely. He suggests a dogmatic emphasis on eliminating all occurrences of minor injury is unlikely to secure that objective. Dekker (2017) points outs: 'If we want to move toward zero losses (and particularly zero fatalities and life-changing injuries), then we should not be obsessed with the "holes" (or minor injuries) that show up in safety management systems. Instead, we should study success. We need to form a deep understanding of how things actually go right, and then enhance the system's capacity to make even more things go right' (p. 105).

9.6   Fostering Collaboration with Regard to Work Health and Safety

The construction industry's specialization and fragmentation make it challenging to achieve collaboration on WHS (and other project goals). Fragmentation in construction projects has been described as being both horizontal and vertical (Fellows and Liu 2012). Horizontal fragmentation describes a reliance on many actors (individuals, organizations, business units) to carry out different functions at the same stage of a construction project. Vertical fragmentation describes how different stages of a construction project involve contributions from different functional actors.

Fellows and Liu (2012) argue that vertical fragmentation (for example, the interaction between a commissioning client, the designers, and the constructors) and horizontal elements (such as interfaces between subcontractors) can create challenges for coordination and alignment. These challenges have the potential to impact WHS. The situation is further complicated by the fact that project participants are engaged at different times, and are often physically separated from one another, making communication and coordination more difficult.

However, the performance of projects (including their WHS performance) depends largely upon how well boundary activities are planned and managed, and the extent to which these boundaries allow information flow, knowledge sharing, and learning.

Failure to adequately plan and manage boundary activities can result in poor safety outcomes. For example, Priemus and Ale (2010) describe how fragmentation in designing and delivering a mixed-use development of commercial, residential, and recreational facilities in the Netherlands contributed to a major structural safety failure. To meet a tight deadline, the project was divided into three parts, each requiring a separate building permit. Further, responsibility for delivery was split between two developers, two building agencies (without a senior structural engineer), three architects' firms (with no consistent overall and final responsibility), one main contractor, and around 50 subcontractors. Priemus and Ale (2011) describe how the coherence of decision making was compromised, communication was ineffective, and project monitoring control systems failed.

Integration is a recurring theme throughout this book. The construction industry's reliance on competitive tendering means relationships are usually transactional (contractual) and 'arms-length'. Further, competitive tendering increases the frequency with which project teams change as different constellations of actors are formed and then disbanded. This instability limits opportunities for learning from experience (Briscoe and Dainty 2005). Gadde and Dubois (2010) argue that it takes time to foster trust, a shared culture, and a mutual orientation to particular issues. It is a process that runs beyond the duration of a single project. Notwithstanding these structural challenges, the advantages associated with fostering a more integrated approach to managing WHS, both horizontally and vertically, are substantial. This can be seen in effective client leadership in project WHS (Chapter 2), as well as improved safety in design outcomes achieved when people with construction process knowledge are involved in design decision making (Chapter 3).

Project delivery team integration is defined as: 'where different disciplines or organizations with different goals, needs and cultures merge into a single cohesive and mutually supporting unit with collaborative alignment of processes and cultures' (Baiden and Price 2011, p. 129). Project delivery teams are reported to vary in the degree to which they are integrated. When fully integrated, team members form a new team identity and work seamlessly towards mutually beneficial goals. However, when teams are fragmented members continue to pursue individual goals, which may not be consistent with project goals.

Many facets of integration are relevant to effectively managing WHS in construction project teams, including those laid out by Baiden and Price (2011) and reproduced in Table 9.1.

Despite the potential benefits of integration, for WHS as well as other aspects of project performance, it is the case that fragmentation, complexity, and uncertainty in the

Table 9.1 Team integration matrix.

Dimensions of integration	Evidence of practice		
	Full integration	**Partial integration**	**Existence of fragmentation**
Single team focus and objectives	All members have the same focus and work together towards team goals	Members pursue individual objectives but in line with overall project objectives	Members pursue objectives individually without regard to, or in isolation from, others and project objectives
Seamless operation without organizationally defined boundaries	Members form a new single project team with no individual member identity or boundaries and work towards mutually beneficial outcomes	Members operate as individuals but make efforts to collaborate with others on the project to meet individual needs	Continued alignment and affiliation to individual organizations that make up the project team
Unrestricted cross-sharing of information	Availability and access to all project information for all parties involved in the project	Access to project information by a section or sections of the project team	Project information only available to members with responsibility for the section of work
Creation of single and co-located team	A single project team with all members located together in a common office	Individually operated subteams but co-located within a single office environment	Individually located and operated teams
Equitable team relationships, opportunities, and respect for all	All members are treated as having equal and significant professional capability needed on the project	Recognition of professional competence, but mainly in each team member's respective field of expertise	Team members' contribution restricted to their functional project role Team members take decisions individually
'No blame' culture	Collective identification and resolution of problems Collective responsibility for all project outcomes	Team members cooperate in resolving problems, but ultimate responsibility rests with a single party	Individual members are singled out for problems that occur on the project and for undertaking corrective measures

Source: Baiden and Price 2011, p. 132.

construction industry's supply arrangements create significant challenges for sustained performance improvement. This is noted by Harvey et al. (2018) as follows:

> Although construction's orthodox approach to safety may go some way to explaining this decline in safety improvement, progress is also hindered by the project-based nature of the industry which requires a dynamic and decentralized network of organizations. Building for a client means designs are unique, profit margins are low, and work is suited to a loosely coupled and dynamic network of specialist organizations contracted to specific aspects of the build. The temporal nature of work and contracts attracts uncommitted and low-skilled workers; subcontracting limits investment in training and safety management; financial constraints do not allow for contingencies or new ideas; learning is rarely transferred between projects; and the culture of litigation, blame and intolerance stifles progress.
>
> *(p. 108)*

9.7 Considering Construction as a Complex Sociotechnical System

Traditional approaches to managing WHS have been found to be limited for two reasons. First, they narrowly identify incidents as local failures, the 'root cause' of which can be identified and controlled through technological and/or administrative controls, thereby preventing future occurrences of similar events. This approach is reactive in nature. However, Carayon et al. (2015) argue that contemporary risk management approaches, even when applied proactively to anticipate hazards, to quantify risk, and to select appropriate control measures, are 'fundamentally static' in nature; that is so because they focus on identifying and managing risks that are already present or can readily be anticipated as a potential failure (or a chain of failures) in work system components. Carayon et al. (2015) argue that hazards arising at the interface between system elements (for example, as a result of coordination or compatibility issues) may not be picked up using 'root cause' analyses and standard methods for risk identification, assessment, and control.

The inherent limitation of traditional risk management approaches is likely to be particularly acute in the construction project context in which project management processes typically break work down into small chunks: for example, elemental design components, work packages, and activities performed by specialized subcontractors. WHS risks may be managed within these parts, but hazards arising from deficiencies in the interconnectedness between them are much harder to anticipate and manage.

A second limitation, also observed by Carayon et al. (2015) and related to the first limitation, is that traditional WHS management activities have focused on the individual worker and an immediate work task. This traditional lens does not seek to focus on understanding the operation of systemic hazards that arise as a result of the way work is organized, designed, and performed. A similar criticism is levelled against many workplace ergonomics initiatives. Kleiner et al. (2015) suggest such initiatives focus on reductionist 'microergonomic' issues, rather than social, ecological, technical, and organizational factors that contribute to the functionality (or dysfunctionality) of a system of work.

There is a growing recognition that WHS is 'situated, negotiated, generated, and transplanted' in the historical, sociomaterial, and cultural context in which work occurs (Turner and Gray 2009, p. 1260). Thus, the decisions of and interactions between all parties involved in delivering construction projects (including commissioning clients, designers, principal contractors, specialist subcontractors, and suppliers of equipment and materials) potentially can impact the way work is done.

Sociotechnical systems theory is increasingly invoked to understand the performance of complex work systems. A sociotechnical system comprises inter-related subsystems:

- the technology subsystem, which includes equipment, machines, tools, and technology, but also the organization of work; and
- the social subsystem, which includes individuals and teams, and needs for coordination, control, and boundary management (Carayon et al. 2015, p. 550).

Importantly, sociotechnical systems theory acknowledges that elements of the social and technological system interact with one another, as well as with aspects of their external environment. Thus, Kleiner et al. (2015) define a system of work in sociotechnical terms as involving:

(1) two or more persons, interacting with some form of technology (hardware and/or software, procedures);
(2) an internal work environment (both physical and cultural);
(3) an external environment (with nested sub-environments); and
(4) an organizational design and management subsystem.

(Kleiner et al. 2015, p. 641)

In sociotechnical systems theory, the notion of emergence is used to describe system properties arising as a result of dynamic interactions between system components, or with the external environment. The dynamic nature of construction projects may amplify emergent properties. For example, the changing project environment is identified as exerting a substantial effect on WHS. Indeed, construction project personnel believe changing features of external and internal project environments pose greater dangers to workers' health and safety than physical hazards associated with particular activities (Harvey et al. 2018). The latter can be subject to a priori identification and control, while the former are difficult to anticipate and plan for.

The analysis by Harvey et al. (2018) confirmed that individual workers' responses to WHS are shaped by a complex web of inter-related pressures and constraints operating at different levels in the system of work (that is, industry, organizational, and workplace factors). These are illustrated in Figure 9.4.

At an industry level these included:

- the client's conflicting roles and interests;
- clients being under-informed about construction and/or WHS;
- a lack of client buy in or ownership of risk;
- the transient workforce;

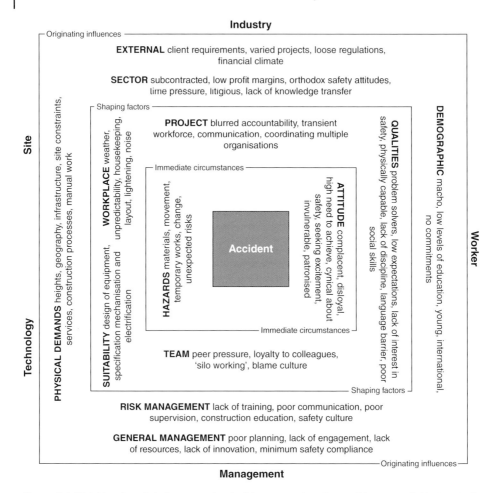

Figure 9.4 Multi-level model of construction incident causation. *Source:* (Harvey et al. 2018, p. 110).

- low levels of trust;
- variability and a lack of knowledge transfer between projects; and
- regulatory deficiencies, including low barriers to entry in the industry.

At an organizational level these included:

- attempts to transfer responsibility for WHS to WHS professionals or subcontractors;
- compartmentalizing and separating WHS from primary activities;
- silo working;
- fear of litigation;
- poorly designed WHS policies (for example, unsuitable rules designed without sensitivity to work tasks);
- deficiencies in planning;
- low levels of integration of WHS into the design stage;
- coordination issues; and
- an acceptance of risk among the workforce.

At a workplace level these included:

- lack of interest and cynicism about WHS in the workforce;
- low levels of worker knowledge about risks; and
- deficiencies in consultative and communication practices.

Further, research in the Australian construction industry supports the multi-level system approach to understanding WHS. This approach suggests that industry, organization, and workplace factors interact in complex and dynamic ways to produce 'vicious cycle' effects. For example, traditional arms-length contractual relationships and an emphasis on price in subcontractor selection, foster low levels of trust in onsite WHS management practices. Clients and principal contractors perceive a need to establish control over the subcontracted workforce through imposing WHS policies and rules. Poor consultation and communication practices impact the suitability and acceptance of rules, which sometimes cannot be followed if subcontractors are also to make a profit and complete work on time. In the face of conflicting client priorities and an emphasis on timely completion, workers become cynical and sometimes break WHS rules in the interests of 'getting the job done'. Trust is further diminished as clients and principal contractors seek to maintain control through enforcement and punitive management processes.

This example of a 'vicious cycle' shows how the way that work is organized and managed in the construction industry creates systemic forces that can shape the way WHS is practised. The challenge for the construction industry is to understand these systemic forces and make changes to transform vicious cycles into virtuous cycles that are generative of WHS.

Given that factors impacting on WHS operate at different levels within the construction industry, such change will not come easily. Effective change requires a rethink of the organizing principles, governance structures, and management approaches applied to planning, procuring, and delivering construction projects.

In keeping with sociotechnical systems theory, the interaction between industry, organizational, and workplace factors should be considered in the context of the external environment within which the industry operates. The regulatory environment, the economic climate, and the availability of skilled labour are all factors that can shape the way clients, contractors, and suppliers prioritize and manage WHS. These factors also indirectly shape organizations' and workers' responses to WHS management activities. Thus, unless the whole construction industry adopts the same high standards when the industry is booming, clients and principal contractors purchasing services for a single project may be less able to demand higher levels of WHS performance from subcontractors and suppliers. However, under the same conditions, subcontractors or suppliers may be willing to invest in WHS innovations or technologies when clients or principal contractors enter into longer term partnership arrangements with them as a guarantee of future work.

At an industry level, client procurement approaches will shape the degree of overlap between project stages and phases, influencing the extent to which project teams can achieve integration and the way in which interfaces between planning, design, and construction are managed. The choice of contracting strategy establishes the roles and responsibilities of the parties to a construction project, in particular, whether WHS risk is shared or pushed down the contractual chain. In collaborative forms of procurement, sharing responsibility for WHS (and other aspects of project performance) encourages joint problem solving and can support innovation and improvement. Where risk is transferred, a blame culture can develop, which can negatively impact transparency and learning from project events. Tendering practices can impact the way WHS is managed. A heavy emphasis on price in contractor selection can be detrimental to WHS investment and resourcing, and these effects are experienced more intensely by subcontractors. These features of industry operation interact with and shape project-level practices, including the integration of WHS into project decision making, the quality of communication and collaboration between multiple organizations, site management and supervision practices. In turn, project-level factors shape workplace attributes, the physical work site, tools, equipment and materials, and the attitudes and behaviour of workers.

Sociotechnical systems theory provides a potentially useful framework through which WHS can be understood as an emergent property of a complex multi-level industry system. This system comprises a multitude of loosely coupled firms, some of which are large multinational companies, but most of which are small enterprises. The dynamic forces inherent in the industry's operation shape the way work is done and, implicitly, how workers experience WHS.

As one of the authors wrote in the 2013 Editorial of a special issue of *Construction Management and Economics*: 'The expectation that workers be able to work productively without suffering harm as a result of wealth-generating activities is the sign of a mature, responsible and equitable industry' (p. 505). The construction industry is not there yet.

9.8 Concluding Remarks

The construction industry is a 'can do' industry. Construction project teams solve complex problems on a daily basis, making it possible to produce magnificent structures and critical infrastructure. Yet, sadly, too many people die, experience life-changing injury, or become seriously ill, as a result of working in construction. To quote Dr Gerry Ayers[1], following the death of one worker and injury of two others during an incident at a Melbourne construction site on Thursday 8 August 2018: 'I think we just need to slow down and really take a good hard look at how we're building and what people are doing, or perhaps what people aren't doing, just to get the job done. It just shouldn't happen. Not in this day and age' (ABC News 2018).

Discussion and Review Questions

1   Have traditional ways of managing risk and WHS compliance reached the limits of their effectiveness? Are new approaches needed? Why or why not?

---

1 Occupational Health, Safety and Environment Manager of the Construction, Forestry, Maritime, Mining and Energy Union's General and Construction Division, Victorian and Tasmanian Branch.

2 Can the quality of jobs in the construction industry be changed? If so, how?

3 What are the barriers to improving job quality in the construction industry? How can these barriers be overcome?

4 How might sociotechnical systems theory help to improve the way WHS is integrated into construction project management in the future?

References

Abbe, O.O., Harvey, C.M., Ikuma, L.H., and Aghazadeh, F. (2011). Modelling the relationship between occupational stressors, psychosocial/physical symptoms and injuries in the construction industry. *International Journal of Industrial Ergonomics* 41 (2): 106–117.

ABC News (2018). Box Hill crane accident puts worksite safety in the spotlight. www.abc .net.au/news/2018-09-07/crane-accident-box-hill-safety-concerns-after-site-fatality/10210292 (accessed 11 September 2018).

ACT (2012). *Getting Home Safely-Inquiry into Compliance with Work Health and Safety Requirements in the ACT's Construction Industry*. Canberra: Australian Capital Territory (ACT).

Advisory Committee on the Safety of Nuclear Installations (1993). ACSNI Study Group on Human Factors. 3rd report. London: Her Majesty's Stationary Office.

Ahiaga-Dagbui, D.D. and Smith, S.D. (2014). Dealing with construction cost overruns using data mining. *Construction Management and Economics* 32 (7–8): 682–694.

Åkerstedt, T. (2006). Psychosocial stress and impaired sleep. *Scandinavian Journal of Work, Environment & Health* 32 (6): 493–501.

Alavinia, S.M., van Duivenbooden, C., and Burdorf, A. (2007). Influence of work-related factors and individual characteristics on work ability among Dutch construction workers. *Scandinavian Journal of Work, Environment & Health* 33 (5): 351–357.

Albers, J., Russell, S., and Stewart, K. (2004). Concrete leveling techniques – a comparative ergonomic assessment. *Proceedings of the Human Factors and Ergonomics Society Annual Meeting* 48 (12): 1349–1353.

Albers, J., Estill, C., and MacDonald, L. (2005). Identification of ergonomics interventions used to reduce musculoskeletal loading for building installation tasks. *Applied Ergonomics* 36 (4): 427–439.

Aleman, A. and Denys, D. (2014). Mental health: a road map for suicide research and prevention. *Nature* 509: 421–423.

Allaire, Y. and Firsirotu, M.E. (1984). Theories of organisational culture. *Organisation Studies* 5 (3): 193–226.

Alper, S.J. and Karsh, B.T. (2009). A systematic review of safety violations in industry. *Accident Analysis and Prevention* 41: 739–754.

Alvesson, M. (2012). *Understanding Organizational Culture*. London: Sage.

Amagasa, T., Nakayama, T., and Takahashi, Y. (2005). Karojisatsu in Japan: characteristics of 22 cases of work-related suicide. *Journal of Occupational Health* 47 (2): 157–164.

American Institute of Architects (2007). Integrated Project Delivery: A Guide. https://info. aia.org/SiteObjects/files/IPD_Guide_2007.pdf (accessed 12 December 2017).

Integrating Work Health and Safety into Construction Project Management, First Edition.
Helen Lingard and Ron Wakefield.
© 2019 John Wiley & Sons Ltd. Published 2019 by John Wiley & Sons Ltd.

American Society of Civil Engineers (2012). Policy Statement 350 – Construction Site Safety. http://www.asce.org/issues-and-advocacy/public-policy/policy-statement-350---construction-site-safety (accessed 12 December 2018).

Anger, W.K., Elliot, D.L., Bodner, T. et al. (2015). Effectiveness of total worker health interventions. *Journal of Occupational Health Psychology* 20 (2): 226–247.

Ankrah, N., Proverbs, D., and Debrah, Y. (2009). Factors influencing the culture of a construction project organisation. *Engineering, Construction and Architectural Management* 16 (1): 26–47.

Antonsen, S. (2009a). Safety culture and the issue of power. *Safety Science* 47 (2): 183–191.

Antonsen, S. (2009b). Safety culture assessment: a mission impossible? *Journal of Contingencies and Crisis Management* 17 (4): 242–254.

Antonsen, S., Almklov, P., and Fenstad, J. (2008). Reducing the gap between procedures and practice – lessons from a successful safety intervention. *Safety Science Monitor* 12: 1–16.

Arezes, P.M. and Miguel, A.S. (2003). The role of safety culture in safety performance measurement. *Measuring Business Excellence* 7: 20–28.

Arndt, V., Rothenbacher, D., Brenner, H. et al. (1996). Older workers in the construction industry: results of a routine health examination and a five year follow up. *Occupational and Environmental Medicine* 53 (10): 686–691.

Arndt, V., Rothenbacher, D., Daniel, U. et al. (2005). Construction work and risk of occupational disability: a ten year follow up of 14,474 male workers. *Occupational and Environmental Medicine* 62 (8): 559–566.

Ashcroft, D.M., Morecroft, C., Parker, D., and Noyce, P.R. (2005). Safety culture assessment in community pharmacy: development, face validity, and feasibility of the Manchester Patient Safety Assessment Framework. *Quality and Safety in Healthcare* 14 (4): 417–421.

Atkin, M. (2018). The biggest lung disease crisis since asbestos: Our love of stone kitchen benchtops is killing workers, ABC News 7.30, www.abc.net.au/news/2018-10-10/stone-cutting-for-kitchen-benchtops-sparks-silicosis-crisis/10357342 (accessed 11th December 2018).

Atkinson, R. (1999). Project management: cost, time and quality, two best guesses and a phenomenon, it's time to accept other success criteria. *International Journal of Project Management* 17 (6): 337–342.

Atkinson, A.R. and Westall, R. (2010). The relationship between integrated design and construction and safety on construction projects. *Construction Management and Economics* 28 (9): 1007–1017.

Australian Institute of Health and Welfare (2010). *Risk Factors and Participation in Work*. Canberra: Australian Institute of Health and Welfare.

Ayers, G., Culvenor, J.F., Sillitoe, J., and Else, D. (2013). Meaningful and effective consultation and the construction industry of Victoria, Australia. *Construction Management and Economics* 31 (6): 542–567.

Azhar, S. (2011). Building information modelling (BIM): trends, benefits, risks, and challenges for the AEC industry. *Leadership and Management in Engineering* 11 (3): 241–252.

Baccarini, D. (1996). The concept of project complexity – a review. *International Journal of Project Management* 14 (4): 201–204.

Baiche, B., Walliman, N., and Ogden, R. (2006). Compliance with building regulations in England and Wales. *Structural Survey* 24 (4): 279–299.

Baiden, B.K. and Price, A.D.F. (2011). The effect of integration on project delivery team effectiveness. *International Journal of Project Management* 29 (2): 129–136.

Baker, J. (2007). The Report of the BP US Refineries Independent Safety Review Panel. www.ogel.org/article.asp?key=2481 (accessed 12 December 2018).

Bakker, A.B. and Demerouti, E. (2007). The job demands-resources model: state of the art. *Journal of Managerial Psychology* 22: 309–328. https://doi.org/10.1108/02683940710733115.

Ball, M. (2014). *Rebuilding Construction (Routledge Revivals): Economic Change in the British Construction Industry*. Routledge.

Banks, A.P. and Millward, L.J. (2007). Differentiating knowledge in teams: the effect of shared declarative and procedural knowledge on team performance. *Group Dynamics: Theory, Research and Practice* 11 (2): 95–106.

Barling, J., Loughlin, C., and Kelloway, K.E. (2002). Development and test of a model linking safety-specific transformational leadership and occupational safety. *Journal of Applied Psychology* 87 (3): 488–496.

Barling, J., Kelloway, E.K., and Iverson, R.D. (2003). High-quality work, job satisfaction and occupational injuries. *Journal of Applied Psychology* 88 (2): 276–283.

Beal, A. (2007). CDM regulations: 12 years of pain but little gain. *Proceedings of the Institution of Civil Engineers – Civil Engineering* 160: 82–88.

Begg, S., Bright, M., and Harper, C. (2008). *Smoking, Nutrition, Alcohol, Physical Activity and Overweight (SNAPO) Indicators for Health Service Districts*. Brisbane: Queensland Health.

Behm, M. (2005). Linking construction fatalities to the design for construction safety concept. *Safety Science* 43 (8): 589–611.

Behm, M. and Schneller, A. (2013). Application of the Loughborough construction accident causation model: a framework for organizational learning. *Construction Management and Economics* 31 (6): 580–595.

Bellamy, L.J. (2015). Exploring the relationship between major hazard, fatal and non-fatal accidents through outcomes and causes. *Safety Science* 71: 93–103.

BeyondBlue (2014). State of Workplace Mental Health in Australia. www.headsup.org.au/docs/default-source/resources/bl1270-report---tns-the-state-of-mental-health-in-australian-workplaces-hr.pdf?sfvrsn=8 (accessed 12 December 2018).

Bieder, C. and Bourrier, M. (2013). *Trapping Safety into Rules: How Desirable or Avoidable Is Proceduralization?* 27–39. Farnham: Ashgate.

Bjorvatn, B., Sagen, I.M., Oyane, N. et al. (2007). The association between sleep duration, body mass index and metabolic measures in the Hordaland Health Study. *Journal of Sleep Research* 16 (1): 66–76.

Black, C.M. (2008). *Working for a Healthier Tomorrow: Dame Carol Black's Review of the Health of Britain's Working Age Population*. The Stationery Office.

Blewett, V. (2011). *Clarifying Culture*. Canberra: ACT: Safe Work Australia.

Blewett, V., Rainbird, S., Dorrian, J. et al. (2012). Keeping rail on track: preliminary findings on safety culture in Australian rail. *Work* 41 (1): 4230–4236.

Bolt, H., Haslam, R., Gibb, A., and Waterson, P. (2012). Pre-conditioning for success: Characteristics and factors ensuring a safe build for the Olympic Park. Prepared by Loughborough University for the Health and Safety Executive. http://www.hse.gov.uk/researcH/rrpdf/rr955.pdf (accessed 12 December 2018).

Bomel (2001). Improving health and safety in construction: Phase 1: Data collection, review and structuring, Contract Research Report 387/2001. London: Health and Safety Executive. http://www.hse.gov.uk/research/crr_pdf/2001/crr01387.pdf (accessed 12 December 2018).

Bongers, P.M., de Winter, C.R., Kompier, M.A.J., and Hildebrandt, V. (1993). Psychosocial factors at work and musculoskeletal disease. *Scandinavian Journal of Work, Environment & Health* 19: 297–312.

Borsting Jacobsen, H., Caban-Martinez, A., Onyebeke, L. et al. (2013). Construction workers struggle with a high prevalence of mental distress and this is associated with their pain and injuries. *Journal of Occupational and Environmental Medicine* 55 (10): 1197–1204.

Borys, D. (2012a). *Organisational Culture*, Core Body of Knowledge for the Generalist OHS Professional. Tullamarine, Victoria, Australia: Safety Institute of Australia Ltd.

Borys, D. (2012b). The role of safe work method statements in the Australian construction industry. *Safety Science* 50 (2): 210–220.

Boschman, J.S., van der Molen, H.F., Sluiter, J.K., and Frings-Dresen, M.H. (2012). Musculoskeletal disorders among construction workers: a one-year follow-up study. *BMC Musculoskeletal Disorders* 13 (1): 196.

Boschman, J.S., Van der Molen, H.F., Sluiter, J.K., and Frings-Dresen, M.H.W. (2013). Psychosocial work environment and mental health among construction workers. *Applied Ergonomics* 44 (5): 748–755.

Boschman, J.S., van der Molen, H.F., Frings-Dresen, M.H.W., and Sluiter, J.K. (2014). The impact of common mental disorders on work ability in mentally and physically demanding construction work. *International Archives of Occupational and Environmental Health* 87 (1): 51–59.

Bowen, P., Edwards, P., Lingard, H., and Cattell, K. (2013a). Workplace stress, stress effects, and coping mechanisms in the construction industry. *Journal of Construction Engineering and Management* 140 (33): 04013059.

Bowen, P.A., Edwards, P.J., and Lingard, H. (2013b). Workplace stress among construction professionals in South Africa: the role of harassment and discrimination. *Engineering, Construction and Architectural Management* 20 (6): 620–635.

Brace, C., Gibb, A., Pendlebury, M., and Bust, P. (2009). Health and safety in the construction industry: Underlying causes of construction fatal accidents – External research. Secretary of State for Work and Pensions, Inquiry into the underlying causes of construction fatal accidents, Loughborough University, Loughborough. http://www.hse.gov.uk/construction/resources/phase2ext.pdf (accessed 12 December 2018).

Brandt, M., Madeleine, P., Ajslev, J.Z.N. et al. (2015). Participatory intervention with objectively measured physical risk factors for musculoskeletal disorders in the construction industry: study protocol for a cluster randomized controlled trial. *BMC Musculoskeletal Disorders* 16 (1): 302.

Breggin, L.K. and Carothers, L. (2006). Governing uncertainty: the nanotechnology environmental, health, and safety challenge. *Columbia Journal of Environmental Law* 31: 285–329.

Brenner, H. and Ahern, W. (2000). Sickness absence and early retirement on health grounds in the construction industry in Ireland. *Occupational and Environmental Medicine* 57 (9): 615–620.

Bresnen, M. and Marshall, N. (2000). Partnering in construction: a critical review of issues, problems and dilemmas. *Construction Management and Economics* 18 (2): 229–237.

Briscoe, G. and Dainty, A. (2005). Construction supply chain integration: an elusive goal? *Supply Chain Management: An International Journal* 10 (4): 319–326.

Brumberger, E.R. (2007a). Visual communication in the workplace: a survey of practice. *Technical Communication Quarterly* 16: 369–395.

Brumberger, E.R. (2007b). Making the strange familiar: a pedagogical exploration of visual thinking. *Journal of Business and Technical Communication* 21: 376–401.

Buchholz, B., Paquet, V., Wellman, H., and Forde, M. (2003). Quantification of ergonomic hazards for ironworkers performing concrete reinforcement tasks during heavy highway construction. *AIHA Journal* 64 (2): 243–250.

Buik, C. and Richards, P. (2015). Workplace health in Australia: BUPA Benchmark Survey 2015. www.bupa.com.au/staticfiles/BupaP3/pdfs/2015%20Bupa%20Workplace% 20Health%20Report.pdf (accessed 12 December 2018).

Burdorf, A., Windhorst, J., van der Beek, A.J. et al. (2007). The effects of mechanised equipment on physical load among road workers and floor layers in the construction industry. *International Journal of Industrial Ergonomics* 37 (2): 133–143.

Bureau of Labor Statistics (2012). Occupational Injury and Illness Classification Manual. http://www.bls.gov/iif/oiics_manual_2010.pdf (accessed 12 December 2018).

Bureau of Labor Statistics (2017). Injuries, Illnesses, and Fatalities: Occupational Safety and Health Definitions. https://www.bls.gov/iif/oshdef.htm (accessed 12 December 2018).

Bureau of Labor Statistics (2018), https://www.bls.gov/home.htm (accessed 27 November 2018).

Burke, M.J., Sarpy, S.A., Smith-Crowe, K. et al. (2006). Relative effectiveness of worker safety and health training methods. *American Journal of Public Health* 96 (2): 315–324.

Burns, C., Mearns, K.Y., and McGeorge, P. (2006). Explicit and implicit trust within safety culture. *Risk Analysis* 26 (5): 1139–1150.

Burrell, G. and Morgan, G. (1994). *Sociological Paradigms and Organisational Analysis*. Aldershot: Arena.

Burt, C.D.B., Gladstone, K.L., and Grieve, K.R. (1998). Development of the considerate and responsible employee (CARE) scale. *Work & Stress* 12 (4): 362–369.

Burt, C.D.B., Sepie, B., and McFadden, G. (2008). The development of a considerate and responsible safety attitude in work teams. *Safety Science* 46: 79–91.

Burton, N.W. and Turrell, G. (2000). Occupation, hours worked, and leisure-time physical activity. *Preventive Medicine* 31 (6): 673–681.

Bustillo, R.M., Fernández-Macías, E., Antón, J.I., and Esteve, F. (2009). *Indicators of Job Quality in the European Union*. Brussels: Policy Department Economic and Scientific Policy European Parliament.

Cadieux, J., Roy, M., and Desmarais, L. (2006). A preliminary validation of a new measure of occupational health and safety. *Journal of Safety Research* 37: 413–419.

Campbell, F. (2006). *Occupational Stress in the Construction Industry*. Berkshire, UK: Chartered Institute of Building.

Cancer Council Australia (2017). Silica dust. Occupational Cancer Risk Series, fact sheet, Cancer Council Australia. www.cancer.org.au/content/Preventing%20cancer/ workplace/2017/2017-10-09-Silica-Factsheet-Final.pdf (accessed 20 March 2018).

Cannon-Bowers, J.A., Salas, E., and Converse, S.A. (1993). Shared mental models in expert team decision making. In: *Individual and Group Decision Making: Current Issues* (ed. J.J. Castellan Jr.), 221–246. Hillsdale, NJ: LEA.

Carayon, P., Smith, M.J., and Haims, M.C. (1999). Work organization, job stress, and work-related musculoskeletal disorders. *Human Factors: The Journal of the Human Factors and Ergonomics Society* 41 (4): 644–663.

Carayon, P., Hancock, P., Leveson, N. et al. (2015). Advancing a sociotechnical systems approach to workplace safety – developing the conceptual framework. *Ergonomics* 58 (4): 548–564.

Carim, G.C., Saurin, T.A., Havinga, J. et al. (2016). Using a procedure doesn't mean following it: a cognitive systems approach to how a cockpit manages emergencies. *Safety Science* 89: 147–157.

Carson J Spencer Foundation & National Action Alliance for Suicide Prevention (2015). *A Construction Industry Blueprint: Suicide Prevention in the Workplace*. Denver, CO: Carson J Spencer Foundation.

Centre for the Protection of Workers' Rights (2013). *The Construction Chart Book*, 5e. Silver Spring, MD, USA: Centre for the protection of Workers' Rights.

Centres for Disease Control and Prevention (2008). National Construction Agenda for Occupational Safety and Health Research and Practice in the U.S. Construction Sector. http://www.cdc.gov/nora/comment/agendas/construction (accessed 12 December 2018).

Chan, D.W.M., Chan, A.P.C., and Choi, T.N.Y. (2010). An empirical survey of the benefits of implementing pay for safety scheme (PFSS) in the Hong Kong construction industry. *Journal of Safety Research* 41 (5): 433–443.

Choi, T.N.Y., Chan, D.W.M., and Chan, A.P.C. (2011). Perceived benefits of applying Pay for Safety Scheme (PFSS) in construction: a factor analysis approach. *Safety Science* 49 (6): 813–823.

Choi, T.N.Y., Chan, D.W.M., and Chan, A.P.C. (2012). Potential difficulties in applying the Pay for Safety Scheme (PFSS) in construction projects. *Accident Analysis and Prevention* 48: 145–155.

Christensen, E.W. and Gordon, G.G. (1999). An exploration of industry, culture and revenue growth. *Organisation Studies* 20 (3): 397–422.

Christian, M., Bradley, J., Wallace, C., and Burke, M. (2009). Workplace safety: a meta-analysis of the roles of person and situation factors. *Journal of Applied Psychology* 94 (5): 1103–1127.

Chu, C., Driscoll, T., and Dwyer, S. (1997). The health-promoting workplace: an integrative perspective. *Australian and New Zealand Journal of Public Health* 21 (4): 377–385.

Claessen, H., Arndt, V., Drath, C., and Brenner, H. (2009). Overweight, obesity and risk of work disability: a cohort study of construction workers in Germany. *Occupational and Environmental Medicine* 66 (6): 402–409.

Clarke, S. (1998). Organisational factors affecting the incident reporting of train drivers. *Work & Stress* 12 (1): 6–16.

Clarke, S. and Ward, K. (2006). The role of leader influence tactics and safety climate in engaging employees' safety participation. *Risk Analysis* 26 (5): 1175–1185.

Cole, K. (2016). Investigating best practice to prevent illness and disease in tunnel construction workers. www.churchilltrust.com.au/media/fellows/Cole_K_2016_Prevention_of_illness_in_tunnel_construction_workers.pdf (accessed 7 March 2018).

Collett, J. (2014) Slogging on until you're 70? Melbourne: The Age. www.theage.com.au/money/saving/slogging-on-until-youre-70-20140419-36xox.html (accessed 12 December 2018).

Comai, A. (2015). Decision-making support: the role of data visualization in analyzing complex systems. *World Future Review* 6 (4): 477–484.

Conchie, S.M. and Burns, C. (2008). Trust and risk communication in high-risk organisations: a test of principles from social risk research. *Risk Analysis* 28 (1): 141–149.

Conchie, S.M. and Burns, C. (2009). Improving occupational safety: using a trusted information source to communicate about risk. *Journal of Risk Research* 12 (1): 13–25.

Conchie, S.M. and Donald, I.J. (2008). The functions and development of safety-specific trust and distrust. *Safety Science* 46 (1): 92–103.

Conchie, S.M., Taylor, P.J., and Charlton, A. (2011). Trust and distrust in safety leadership: mirror reflections? *Safety Science* 49 (8–9): 1208–1214.

Constructing Better Health (2018. About Constructing Better Health (CBH). https://www.cbhscheme.com/About-CBH (accessed 27 November 2018).

Cooke, T. and Lingard, H. (2011). A retrospective analysis of work-related deaths in the Australian construction industry. In: *Proceedings 27th Annual ARCOM Conference*, Bristol, 5–7 September (ed. C. Egbu and E.C.W. Lou), 279–288. Reading: Association of Researchers in Construction Management.

Cooke, D.L. and Rohleder, T.R. (2006). Learning from incidents: from normal accidents to high reliability. *System Dynamics Review* 22: 213–239.

Cooke, T., Lingard, H., Blismas, N., and Stranieri, A. (2008). ToolSHeD: the development and evaluation of a decision support tool for health and safety in construction design. *Engineering, Construction and Architectural Management* 15 (4): 336–351.

Cooper, M.D. and Phillips, R.A. (2004). Exploratory analysis of the safety climate and safety behaviour relationship. *Journal of Safety Research* 35 (5): 497–512.

Cox, S. and Flin, R. (1998). Safety culture: Philosopher's stone or man of straw? *Work & Stress* 12 (3): 189–201.

Crossrail Limited (2017). Best Practice Guide: Air Quality, Crossrail Limited. https://learninglegacy.crossrail.co.uk/wp-content/uploads/2017/03/HS17_BPG_AirQuality.pdf (accessed 21 March 2018).

Cullen, D. (1990). The Public Inquiry into the Piper Alpha Disaster. London: Her Majesty's Stationary Office. http://www.hse.gov.uk/offshore/piper-alpha-disaster-public-inquiry.htm (accessed 12 December 2018).

Dahl, O. (2013). Safety compliance in a highly regulated environment: a case study of workers' knowledge of rules and procedures within the petroleum industry. *Safety Science* 60: 185–195.

Dainty, A.R.J. and Lingard, H. (2006). Indirect discrimination in construction organisations and the impact on women's careers. ASCE. *Journal of Management in Engineering* 22 (3): 108–118.

Dale, A.M., Ryan, D., Welch, L. et al. (2015). Comparison of musculoskeletal disorder health claims between construction floor layers and a general working population. *Occupational and Environmental Medicine* 72 (1): 15–20.

Dale, A.M., Jaegers, L., Welch, L. et al. (2016). Evaluation of a participatory ergonomics intervention in small commercial construction firms. *American Journal of Industrial Medicine* 59 (6): 465–475.

De Boer, A.G., Burdorf, A., van Duivenbooden, C., and Frings-Dresen, M.H. (2007). The effect of individual counselling and education on work ability and disability pension: a prospective intervention study in the construction industry. *Occupational and Environmental Medicine* 64 (12): 792–797.

De Wind, A., Geuskens, G.A., Reeuwijk, K.G. et al. (2013). Pathways through which health influences early retirement: a qualitative study. *BMC Public Health* 13: https://doi.org/10.1186/1471-2458-13-292.

De Zwart, C.H., Frings-Dresen, M.H.W., and Van Duivenbooden, J.C. (1999). Senior workers in the Dutch construction industry: a search for age-related work and health issues. *Experimental Aging Research* 25 (4): 385–391.

Deeney, C. and O'Sullivan, L. (2009). Work related psychosocial risks and musculoskeletal disorders: potential risk factors, causation and evaluation methods. *Work* 34 (2): 239–248.

DeJoy, D.M., Wilson, M.G., Vandenberg, R.J. et al. (2010). Assessing the impact of a healthy work organisation intervention. *Journal of Occupational and Organisational Psychology* 83 (1): 139–165.

Dekker, S.W. (2002). Reconstructing human contributions to accidents: the new view on error and performance. *Journal of Safety Research* 33 (3): 371–385.

Dekker, S. (2003). Failure to adapt of adaptations that fail: contrasting models on procedures and safety. *Applied Ergonomics* 34: 233–238.

Dekker, S. (2005). *Ten Questions about Human Error: A New View of Human Factors and System Safety*. Boca Raton, FL: CRC Press.

Dekker, S. (2008). *Just Culture: Balancing Safety and Accountability*. Burlington, VT: Ashgate Publishing.

Dekker, S.W. (2014). The bureaucratization of safety. *Safety Science* 70: 348–357.

Dekker, S. (2017). Zero vision: enlightenment and new religion. *Policy and Practice in Health and Safety* 15 (2): 101–107.

Dekker, S. and Pitzer, C. (2016). Examining the asymptote in safety progress: a literature review. *International Journal of Occupational Safety and Ergonomics* 22 (1): 57–65.

Dement, J., Ringen, K., Welch, L. et al. (2005). Surveillance of hearing loss among older construction and trade workers at Department of Energy nuclear sites. *American Journal of Industrial Medicine* 48 (5): 348–358.

Demerouti, E., Bakker, A.B., Nachreiner, F., and Schaufeli, W.B. (2001). The job demands–resources model of burnout. *Journal of Applied Psychology* 86: 499–512. https://doi.org/10.1037/0021-9010.86.3.499.

Demers, C. (2007). *Organisational Change Theories: A Synthesis*. Los Angeles: Sage.

Department of Treasury and Finance (2009). *In Pursuit of Additional Value: A Benchmarking Study into Alliancing in the Australian Public Sector*. Melbourne, Victoria: Victorian Government.

van der Molen, H.F., Grouwstra, R., Kuijer, P.P.F. et al. (2004). Efficacy of adjusting working height and mechanizing of transport on physical work demands and local discomfort in construction work. *Ergonomics* 47 (7): 772–783.

van der Molen, H.F., Sluiter, J.K., and Frings-Dresen, M.H. (2005a). Behavioural change phases of different stakeholders involved in the implementation process of ergonomics measures in bricklaying. *Applied Ergonomics* 36 (4): 449–459.

van der Molen, H.F., Sluiter, J.K., Hulshof, C.T. et al. (2005b). Implementation of participatory ergonomics intervention in construction companies. *Scandinavian Journal of Work, Environment & Health* 31: 191–204.

van der Molen, H.F., Koningsveld, E., Haslam, R., and Gibb, A. (2005c). Editorial.). Ergonomics in building and construction: time for implementation. *Applied Ergonomics* 36: 387–389.

Devine, C.M., Jastran, M., Jabs, J. et al. (2006). 'A lot of sacrifices': work–family spillover and the food choice coping strategies of low-wage employed parents. *Social Science & Medicine* 63 (10): 2591–2603.

Donaghy, R. (2009). One Death is too Many: Inquiry into the Underlying Causes of Construction Fatal Accidents, Report to the Secretary of State for Work and Pensions. Richmond. UK: Office of Public Information. https://assets.publishing.service.gov.uk/government/uploads/system/uploads/attachment_data/file/228876/7657.pdf (accessed 13 December 2018).

Doran, C. and Ling, R. (2014). The economic cost of suicide and suicide behaviour in the NSW construction industry and the impact of MATES in Construction suicide prevention strategy in reducing this cost. http://micaus.bpndw46jvgfycmdxu.maxcdn-edge.com/wp-content/uploads/2016/03/2014-Economic-cost-of-suicide-in-NSW.pdf (accessed 13 December 2018).

Driscoll, T.R., Harrison, J.E., Bradley, C., and Newson, R.S. (2008). The role of design issues in work-related fatal injury in Australia. *Journal of Safety Research* 39 (2): 209–214.

Driskell, J.E., Salas, E., and Johnston, J. (1999). Does stress lead to a loss of team perspective? *Group Dynamics: Theory, Research, and Practice* 3 (4): 291–302.

Du Plessis, K., Cronin, D., Corney, T., and Green, E. (2013). Australian blue-collar men's health and wellbeing: contextual issues for workplace health promotion interventions. *Health Promotion Practice* 14 (5): 715–720.

Dubois, A. and Gadde, L.E. (2002). The construction industry as a loosely coupled system: implications for productivity and innovation. *Construction Management and Economics* 20 (7): 621–631.

Dyreborg, J. (2009). The causal relation between lead and lag indicators. *Safety Science* 47: 474–4475.

Eatough, E.M., Way, J.D., and Chang, C.H. (2012). Understanding the link between psychosocial work stressors and work-related musculoskeletal complaints. *Applied Ergonomics* 43 (3): 554–563.

Eban, G. (2016). Major construction projects at airports: client leadership of health and safety. *Journal of Airport Management* 10 (2): 131–137.

Eisenberger, R., Fasolo, P., and LaMastro, D. (1990). Perceived organisational support and employee diligence, commitment and innovation. *Journal of Applied Psychology* 75 (1): 51–59.

Engholm, G. and Holmström, E. (2005). Dose-response associations between musculoskeletal disorders and physical and psychosocial factors among construction workers. *Scandinavian Journal of Work, Environment & Health* 31: 57–67.

English, D. and Branaghan, R.J. (2012). An empirically derived taxonomy of pilot violation behaviour. *Safety Science* 50: 199–209.

Entzel, P., Albers, J., and Welch, L. (2007). Best practices for preventing musculoskeletal disorders in masonry: stakeholder perspectives. *Applied Ergonomics* 38 (5): 557–566.

Environment Transport and Works Bureau (2000). *Chapter 12 – Pay for Safety Scheme*. Hong Kong: Hong Kong SAR Government.

Estrada, F.C.R. and Davis, L.S. (2015). Improving visual communication of science through the incorporation of graphic design theories and practices into science communication. *Science Communication* 37 (1): 140–148.

Ettner, S.L. and Grzywacz, J. (2001). Workers' perceptions of how jobs affect health: a social ecological perspective. *Journal of Occupational Health Psychology* 6 (2): 101–113.

European Agency for Safety and Health at Work (EU-OSHA) (2012). Drivers and barriers for psychosocial risk management: an analysis of findings of the of the European Survey of Enterprises on New and Emerging Risks (ESENER). Luxembourg: Publications Office of the European Union. https://osha.europa.eu/en/tools-and-publications/publications/reports/drivers-barriers-psychosocial-risk-management-esener (accessed 13 December 2018).

European Federation of Foundation Contractors (2015). Breaking Down of Concrete Piles. https://www.effc.org/content/uploads/2015/12/Breaking_Down_of_Piles_May2015.pdf (accessed 13 December 2018).

Fadier, E. and De la Garza, C. (2006). Safety design: towards a new philosophy. *Safety Science* 44: 55–73.

Fang, D., Yang, C., and Wong, L. (2006). Safety climate in construction industry: a case study in Hong Kong. *Journal of Construction Engineering and Management* 132: 573–584.

Fellows, R. and Liu, A.M. (2012). Managing organizational interfaces in engineering construction projects: addressing fragmentation and boundary issues across multiple interfaces. *Construction Management and Economics* 30 (8): 653–671.

Fellows, R. and Liu, A.M.M. (2013). Use and misuse of the concept of culture. *Construction Management and Economics* 31 (5): 401–422.

Filho, A.P.G., Andrade, J.C.S., and Marinho, M.D.O. (2010). A safety culture maturity model for petrochemical companies in Brazil. *Safety Science* 48 (5): 615–624.

Findley, M., Smith, S., Gorski, J., and O'Neil, M. (2007). Safety climate differences among job positions in a nuclear decommissioning and demolition industry: employees' self-reported safety attitudes and perceptions. *Safety Science* 45: 875–889.

Fiore, S.M., Salas, E., Cuevas, H.M., and Bowers, C.A. (2003). Distributed coordination space: toward a theory of distributed team process and performance. *Theoretical Issues in Ergonomics Science* 4: 340–364.

Fisher, D.M., Bell, S.T., Dierdorff, E.C., and Belohlav, J.A. (2012). Facet personality and surface-level diversity as team mental model antecedents: implications for implicit coordination. *Journal of Applied Psychology* 97 (4): 825–841.

Flin, R. and Yule, S. (2004). Leadership for safety: industrial experience. *Quality and Safety in Health Care* 13: 45–51.

Flin, R., Mearns, K., O'Connor, P., and Bryden, R. (2000). Measuring safety climate: identifying the common features. *Safety Science* 34 (1–3): 177–192.

Forde, M.S., Punnett, L., and Wegman, D.H. (2005). Prevalence of musculoskeletal disorders in union ironworkers. *Journal of Occupational and Environmental Hygiene* 2 (4): 203–212.

Franz, B.W., Leicht, R.M., and Riley, D.R. (2013). Project impacts of specialty mechanical contractor design involvement in the health care industry: comparative case study. *Journal of Construction Engineering and Management* 139 (9): 1091–1097.

Frick, K. (2011). Worker influence on voluntary OHS management systems – a review of its ends and means. *Safety Science* 49 (7): 974–987.

Frick, K. and Wren, J. (2000). Reviewing occupational health and safety management – multiple roots, diverse perspectives and ambiguous outcomes. In: *Systematic Occupational Health and Safety Management: Perspectives on an International Development* (ed. K. Frick, P.L. Jensen, M. Quinlan and T. Wilthagen), 17–42. Amsterdam: Pergamon.

Fridner, A., Belkic, K., Marini, M. et al. (2009). Survey on recent suicidal ideation among female university hospital physicians in Sweden and Italy (the HOUPE study): cross-sectional associations with work stressors. *Gender Medicine* 6 (1): 314–328.

Fritz, C. and Sonnentag, S. (2005). Recovery, health, and job performance: effects of weekend experiences. *Journal of Occupational Health Psychology* 10 (3): 187.

Frone, M.R., Russell, M., and Cooper, M.L. (1997). Relation of work–family conflict to health outcomes: a four-year longitudinal study of employed parents. *Journal of Occupational & Organizational Psychology* 70 (4): 325–335.

Frost, S. and Harding, A.-H. (2015). The effect of wearer stubble on the protection given by Filtering Facepieces Class 3 (FFP3) and Half Masks. www.hse.gov.uk/research/rrhtm/rr1052.htm (accessed 15 February 2018).

Fruhen, L.S., Mearns, K.J., Flin, R.H., and Kirwan, B. (2013). From the surface to the underlying meaning-an analysis of senior managers' safety culture perceptions. *Safety Science* 57: 326–334.

Fucks, I. and Dien, Y. (2013). 'No rule, no use'? The effects of over-proceduralization. In: *Trapping Safety into Rules: How Desirable or Avoidable Is Proceduralization?* (ed. C. Bieder and M. Bourrier), 27–39. Farnham: Ashgate.

Gadde, L.E. and Dubois, A. (2010). Partnering in the construction industry – problems and opportunities. *Journal of Purchasing and Supply Management* 16 (4): 254–263.

Gale, A. and Cartwright, S. (1995). Women in project management: entry into a male domain?: a discussion on gender and organisational culture – part 1. *Leadership & Organisation Development Journal* 16 (2): 3–8.

Gallagher, C., Underhill, E., and Rimmer, M. (2003). Occupational safety and health management systems in Australia: barriers to success. *Policy and Practice in Health and Safety* 1 (2): 67–81.

Gambatese, J. and Hinze, J. (1999). Addressing construction worker safety in the design phase: designing for construction worker safety. *Automation in Construction* 8 (6): 643–649.

Gambatese, J.A., Behm, M., and Hinze, J.W. (2005). Viability of designing for construction worker safety. *Journal of Construction Engineering and Management* 131 (9): 1029–1036.

Gambatese, J.A., Behm, M., and Rajendran, S. (2008). Design's role in construction accident causality and prevention: perspectives from an expert panel. *Safety Science* 46 (4): 675–691.

Gangolells, M., Casals, M., Forcada, N. et al. (2010). Mitigating construction safety risks using prevention through design. *Journal of Safety Research* 41 (2): 107–121.

Gangwisch, J.E., Malaspina, D., Boden-Albala, B., and Heymsfield, S.B. (2005). Inadequate sleep as a risk factor for obesity: analyses of the NHANES I. *Sleep* 28 (10): 1289–1296.

Gangwisch, J.E., Heymsfield, S.B., Boden-Albala, B. et al. (2006). Short sleep duration as a risk factor for hypertension: analyses of the first National Health and Nutrition Examination Survey. *Hypertension* 47 (5): 833–839.

Garrett, J.W. and Teizer, J. (2009). Human factors analysis classification system relating to human error awareness taxonomy in construction safety. *Journal of Construction Engineering and Management* 135 (8): 754–763.

Gaupset, S. (2000). The Norwegian internal control reform – an unrealized potential? In: *Systematic Occupational Health and Safety Management: Perspectives on an International Development* (ed. K. Frick, P.L. Jensen, M. Quinlan and T. Wilthagen), 329–348. Amsterdam: Pergamon.

Geertz, C. (1973). *The Interpretation of Cultures: Selected Essays.* London: Hutchinson & Co.

Gervais, M. (2003). Good management practice as a means of preventing back disorders in the construction sector. *Safety Science* 41 (1): 77–88.

Geurts, S.A. and Sonnentag, S. (2006). Recovery as an explanatory mechanism in the relation between acute stress reactions and chronic health impairment. *Scandinavian Journal of Work, Environment & Health* 32: 482–492.

Gherardi, S., Nicolini, D., and Odella, F. (1998). What do you mean by safety? Conflicting perspectives on accident causation and safety management in a construction firm. *Journal of Contingencies and Crisis Management* 6 (4): 202–213.

Gherardi, S., and Nicolini, D. (2002). Learning the trade: A culture of safety in practice. *Organization* 9 (2): 191–223.

Gibb, A., Haslam, R., Pavitt, T., and Horne, K. (2007). Designing for health – reducing occupational health risks in bored pile operations. *Construction Information Quarterly, Special Issue: Health and Safety* 9 (3): 113–123.

Gibb, A., Lingard, H., Behm, M., and Cooke, T. (2014). Construction accident causality: learning from different countries and consequence. *Construction Management and Economics* 32: 446–459.

Gillen, M., Baltz, D., Gassel, M. et al. (2002). Perceived safety climate, job demands and co-worker support among union and non-union injured construction workers. *Journal of Safety Research* 33: 33–51.

Glendon, A.I. and Litherland, D.K. (2001). Safety climate factors, group differences and safety behaviour in road construction. *Safety Science* 39 (3): 157–188.

Glendon, I. and Stanton, N.A. (2000). Perspectives on safety culture. *Safety Science* 34: 193–214.

Glendon, A.I., Clarke, S.G., and McKenna, E.F. (2006). *Human Safety and Risk Management.* Boca Raton: CRC Press.

Goldenhar, L., Williams, L.J., and Swanson, N.G. (2003). Modelling relationships between job stressors and injury and near-miss outcomes for construction labourers. *Work & Stress* 17 (3): 218–240.

Goldsheyder, D., Weiner, S.S., Nordin, M., and Hiebert, R. (2004). Musculoskeletal symptom survey among cement and concrete workers. *Work* 23 (2): 111–121.

Gordon, G.G. (1991). Industry determinants of organisational culture. *Academy of Management Review* 16 (2): 396–415.

Gordon, R. P. (1998). The contribution of human factors to accidents in the offshore oil industry. Reliability Engineering & System Safety 61 (1-2): 95–108.

Gottlieb, D.J., Redline, S., Nieto, F.J. et al. (2006). Association of usual sleep duration with hypertension: the Sleep Heart Health Study. *Sleep* 29 (8): 1009–1014.

Gram, B., Holtermann, A., Sogaard, K., and Sjogaard, G. (2012). Effect of individualized worksite exercise training on aerobic capacity and muscle strength among construction workers – a randomized controlled intervention study. *Scandinavian Journal of Work, Environment & Health* 38 (5): 467–475.

Grandey, A.A., Cordeiro, B.L., and Michael, J.H. (2007). Work-family supportiveness organizational perceptions: important for the well-being of male blue-collar hourly workers? *Journal of Vocational Behavior* 71 (3): 460–478.

Grant, L. and Kinman, G. (2013). The importance of emotional resilience for staff and students in the 'helping' professions: developing an emotional curriculum. The Higher Education Academy: Health and Social Care. https://www.heacademy.ac.uk/system/files/emotional_resilience_louise_grant_march_2014_0.pdf (accessed 13 December 2018).

Greenhaus, J.H. and Beutell, N.J. (1985). Sources of conflict between work and family roles. *Academy of Management Review* 10 (1): 76–88.

Greenhaus, J.H., Collins, K.M., and Shaw, J.D. (2003). The relation between work–family balance and quality of life. *Journal of Vocational Behavior* 63 (3): 510–531.

Griffin, M.A. and Neal, A. (2000). Perceptions of safety at work: a framework for linking safety climate to safety performance, knowledge, and motivation. *Journal of Occupational Health Psychology* 5 (3): 347–358.

Groeneveld, I.F., Proper, K.I., van der Beek, A.J., and van Mechelen, W. (2010). Sustained body weight reduction by an individual-based lifestyle intervention for workers in the construction industry at risk for cardiovascular disease: results of a randomized controlled trial. *Preventive Medicine* 51 (3–4): 240–246.

Grzywacz, J.G. and Fuqua, J. (2000). The social ecology of health: leverage points and linkages. *Behavioral Medicine* 26 (3): 101–115.

Gudienė, N., Banaitis, A., Podvezko, V., and Banaitienė, N. (2014). Identification and evaluation of the critical success factors for construction projects in Lithuania: AHP approach. *Journal of Civil Engineering and Management* 20 (3): 350–359.

Guldenmund, F.W. (2000). The nature of safety culture: a review of theory and research. *Safety Science* 34 (1–3): 215–257.

Guldenmund, F.W. (2007). The use of questionnaires in safety culture research – an evaluation. *Safety Science* 45 (6): 723–743.

Gullestrup, J., Lequertier, B., and Martin, G. (2011). MATES in construction: impact of a multimodal, community-based program for suicide prevention in the construction industry. *International Journal of Environmental Research and Public Health* 8 (11): 4180–4196.

Guo, B.H.W., Yiu, T.W., and Gonzalez, V.A. (2016). Predicting safety behaviour in the construction industry: development and test of an integrative model. *Safety Science* 84: 1–11.

Gyekye, S.A. and Salminen, S. (2007). Workplace safety perceptions and perceived organisational support: do supportive perceptions influence safety perceptions? *International Journal of Occupational Safety and Ergonomics* 13 (2): 189–200.

Hale, A.R. (2000). Culture's confusions. *Safety Science* 34 (1–3): 1–14.

Hale, A.R. (2002). Conditions of occurrence of major and minor accidents. Urban myths, deviations and accident scenarios. *Tijdschrift voor toegepaste Arbowetenschap* 15 (3): 34–40. http://www.arbeidshygiene.nl/_uploads/text/file/2002-03_Hale_full%20paper%20trf.pdf, accessed 12 May 2017.

Hale, A.R. (2003). Safety management in production. *Human Factors and Ergonomics in Manufacturing & Service Industries* 13 (3): 185–120.

Hale, A. and Borys, D. (2013a). Working to rule or working safely? Part 1: a state of the art review. *Safety Science* 55: 207–221.

Hale, A. and Borys, D. (2013b). Working to rule or working safely? Part 2: the management of safety rules and procedures. *Safety Science* 255: 22–231.

Hale, A.R. and Hovden, J. (1998). Management and culture: the third age of safety. A review of approaches to organizational aspects of safety, health and environment. In: *Occupational Injury: Risk, Prevention and Intervention* (ed. A.-M. Feyer and A. Williamson), 129–165. London: Taylor & Francis.

Hale, A., Kirwan, B., and Kjellén, U. (2007). Safe by design: where are we now? *Safety Science* 45: 305–327.

Hallowell, M.R., Alexander, D., and Gambatese, J.A. (2017). Energy-based safety risk assessment: does magnitude and intensity of energy predict injury severity? *Construction Management and Economics* 35: 1–14.

Han, S., Saba, F., Lee, S. et al. (2014). Toward an understanding of the impact of production pressure on safety performance in construction operations. *Accident Analysis & Prevention* 68: 106–116.

Hannerz, H., Spangenberg, S., Tüchsen, F., and Albertsen, K. (2005). Disability retirement among former employees at the construction of the Great Belt Link. *Public Health* 119 (4): 301–304.

Hansez, I. and Chmiel, N. (2010). Safety behavior: job demands, job resources, and perceived management commitment to safety. *Journal of Occupational Health Psychology* 15 (3): 267.

Hare, B., Cameron, I., and Duff, A. (2006). Exploring the integration of health and safety with pre-construction planning. *Engineering, Construction and Architectural Management* 13 (5): 438–450.

Harms-Ringdahl, L. (2009). Dimensions in safety indicators. *Safety Science* 47: 481–482.

Harper, D. (2002). Talking about pictures: a case for photo elicitation. *Visual Studies* 17 (1): 13–26.

Harvey, E.J., Waterson, P., and Dainty, A.R. (2018). Beyond ConCA: rethinking causality and construction accidents. *Applied Ergonomics* 73: 108–121.

Haslam, R., Hide, S., Gibb, A., et al. (2003). Causal factors in construction accidents (Research Report 156). Norwich: Health and Safety Executive. http://www.hse.gov.uk/researcH/rrpdf/rr156.pdf (accessed 13 December 2018).

Haukelid, K. (2008). Theories of (safety) culture revisited – an anthropological approach. *Safety Science* 46 (3): 413–426.

Haviland, D. (1996). Some shifts in building design and their implications for design practices and management. *Journal of Architectural and Planning Research* 13 (1): 50–62.

Håvold, J.I. (2010). Safety culture and safety management aboard tankers. *Reliability Engineering & System Safety* 95 (5): 511–519.

Hawton, K., Harriss, L., Simkin, S. et al. (2004). Self-cutting: patient characteristics compared with self-poisoners. *Suicide and Life Threatening Behavior* 34 (3): 199–208.

Health and Safety Commission (2001). *The Ladbroke Grove Rail Inquiry. Part 2 Report.* London: The Rt Hon Lord Cullen, HSE Books.

Health and Safety Executive (1999). *Reducing Error and Influencing Behaviour*. London: HSE Books.

Health and Safety Executive (2003). Causal factors in construction accidents (Research Report 156). Norwich: Health and Safety Executive. http://www.hse.gov.uk/researcH/rrpdf/rr156.pdf (accessed 13 December 2018).

Health and Safety Executive (2005a). A review of safety culture and safety climate literature for the development of the safety culture inspection toolkit (Research Report 367). Norwich: Health and Safety Executive. http://www.hse.gov.uk/research/rrpdf/rr367.pdf (accessed 13 December 2018).

Health and Safety Executive (2005b). Development and validation of the HMRI safety culture inspection toolkit (Research Report 365). Norwich: Health and Safety Executive. http://www.hse.gov.uk/research/rrpdf/rr365.pdf (accessed 13 December 2018).

Health and Safety Executive (2012). Safety culture on the Olympic Park (Research Report 942). Health and Safety Executive. http://www.hse.gov.uk/research/rrpdf/rr942.pdf (accessed 13 December 2018).

Health and Safety Executive (2013). Construction dust: Construction Information Sheet No 36 (Revision 2). www.hse.gov.uk/pubns/cis36.pdf (accessed 13 December 2018).

Health and Safety Executive (2015). Managing health and safety in construction: Construction (Design and Management) Regulations 2015, Guidance on Regulations L153. http://www.hse.gov.uk/pUbns/priced/l153.pdf (accessed 13 December 2018).

Health and Safety Executive (2017). Health and safety in construction sector in Great Britain, 2014/15.

Health and Safety Executive (2018a). Construction Statistics in Great Britain. www.hse.gov.uk/statistics/industry/construction.pdf (accessed 5 December 2018).

Health and Safety Executive (2018b). Health and safety in construction sector in Great Britain, 2017.

Heinrich, H.W. (1931). *Industrial Accident Prevention: A Scientific Approach.* New York: McGraw-Hill Book Company.

Heller, T.S., Hawgood, J.L., and Leo, D.D. (2007). Correlates of suicide in building industry workers. *Archives of Suicide Research* 11 (1): 105–117.

Hess, J.A., Hecker, S., Weinstein, M., and Lunger, M. (2004). A participatory ergonomics intervention to reduce risk factors for low-back disorders in concrete laborers. *Applied Ergonomics* 35 (5): 427–441.

Hess, J.A., Kincl, L., Amasay, T., and Wolfe, P. (2010). Ergonomic evaluation of masons laying concrete masonry units and autoclaved aerated concrete. *Applied ergonomics* 41 (3): 477–483.

Hidden, A. (1989). *Investigation into the Clapham Junction Railway Accident.* London: Her Majesty's Stationary Office.

Hinze, J. (1997). *Construction Safety*. Upper Saddle, NJ: Prentice Hall.

Hinze, J., Thurman, S., and Wehle, A. (2013). Leading indicators of construction safety performance. *Safety Science* 51: 23–28.

Hofmann, D.A. and Stetzer, A. (1998). The role of safety climate and communication in accident interpretation: implications for learning from negative events. *Academy of Management Journal* 41 (6): 644–657.

Hohnen, P. and Hasle, P. (2011). Making work environment auditable – a 'critical case' study of certified occupational health and safety management systems in Denmark. *Safety Science* 49 (7): 1022–1029.

Hollmann, S., Heuer, H., and Schmidt, K.H. (2001). Control at work: a generalized resource factor for the prevention of musculoskeletal symptoms? *Work & Stress* 15 (1): 29–39.

Hollnagel, E. (2010). Extending the scope of the human factor. In: *Safer Complex Industrial Environments* (ed. E. Hollnagel), 37–59. Boca Raton: CRC Press.

Hollnagel, E. (2011). Prologue: the scope of resilience engineering. In: *Resilience Engineering in Practice: A Guidebook* (ed. E. Hollnagel, J. Pariès, D.D. Woods and J. Wreathall). Farnham, UK: Ashgate.

Hollnagel, E. (2014). *Safety-I and Safety-II: The Past and Future of Safety Management.* Farnham, UK: Ashgate.

Hollnagel, E. (2015). Why is work-as-imagined different from work-as-done? In: *Resilient Health Care: The Resilience of Everyday Clinical Work* (ed. R.L. Wears, E. Hollnagel and J. Braithwaite), 249–264. Farnham, UK: Ashgate.

Hollnagel, E. and Amalberti, R. (2001). The emperor's new clothes: or whatever happened to "human error". In: *Proceedings of the 4th International Workshop on Human Error, Safety and Systems Development*, 1–18.

Holmström, E. and Ahlborg, B. (2005). Morning warming-up exercise—effects on musculoskeletal fitness in construction workers. *Applied Ergonomics* 36 (4): 513–519.

Holmström, E.B., Lindell, J., and Moritz, U. (1992). Low back and neck/shoulder pain in construction workers: occupational workload and psychosocial risk factors. Part 1: relationship to low back pain. *Spine* 17 (6): 663–671.

Hong, O. (2005). Hearing loss among operating engineers in American construction industry. *International Archives of Occupational and Environmental Health* 78 (7): 565–574.

van Hooff, M.L., Geurts, S.A., Kompier, M.A., and Taris, T.W. (2007). Workdays, in-between workdays and the weekend: a diary study on effort and recovery. *International Archives of Occupational and Environmental Health* 80 (7): 599–613.

Hoogendoorn, W.E., van Poppel, M.N., Bongers, P.M. et al. (2000). Systematic review of psychosocial factors at work and private life as risk factors for back pain. *Spine* 25 (16): 2114–2125.

Hopkins, A. (2000). *Lessons from Longford: The Esso Gas Plant Explosion*. CCH Australia Ltd.

Hopkins, A. (2006a). Studying Organisational Cultures and their Effects on Safety (Working Paper 44). Australian National University, Canberra: National Research Centre for OHS Regulation. https://openresearch-repository.anu.edu.au/bitstream/1885/43245/2/wp44-StudyingOrganizationalCultures.pdf (accessed 13 December 2018).

Hopkins, A. (2006b). What are we to make of safe behaviour programs? *Safety Science* 44 (7): 583–597.

Hopkins, A. (2009a). Thinking about process safety indicators. *Safety Science* 47: 460–465.

Hopkins, A. (2009b). Reply to comments. *Safety Science* 47: 508–510.

Hopkinson, J., Fox, D., and Lunt, J. (2015). *Development of a Health Risk Management Maturity Index (HeRMMIn) as a Performance Leading Indicator within the Construction Industry*. London: HSE Books.

Houts, P.S., Doak, C.C., Doak, L.G., and Loscalzo, M.J. (2006). The role of pictures in improving health communication: a review of research on attention, comprehension, recall and adherence. *Patient Education and Counseling* 61: 173–190.

Huang, X. and Hinze, J. (2006a). Owner's role in construction safety. *Journal of Construction Engineering and Management* 132: 164–173.

Huang, X. and Hinze, J. (2006b). Owner's role in construction safety. *Journal of Construction Engineering and Management* 132: 174–181.

Huang, G.D., Feuerstein, M., Kop, W.J. et al. (2003). Individual and combined impacts of biomechanical and work organization factors in work-related musculoskeletal symptoms. *American Journal of Industrial Medicine* 43 (5): 495–506.

Hudson, P. (2007). Implementing safety culture in a major multi-national. *Safety Science* 45 (6): 697–722.

Hudson, P.T., Verschuur, W.L.G., Parker, D., et al. (1998). Bending the rules: Managing violation in the workplace. International Conference of the Society of Petroleum Engineers.

Humphrey, S.E., Moon, H., Conlon, D.E., and Hofmann, D.A. (2004). Decision-making and behavior fluidity: how focus on completion and emphasis on safety changes over the course of projects. *Organizational Behavior and Human Decision Processes* 93 (1): 14–27.

Iacuone, D. (2005). Real men are tough guys: hegemonic masculinity and safety in the construction industry. *The Journal of Men's Studies* 13 (2): 247–266.

Inness, T., Turner, N., Barling, J., and Stride, C.B. (2010). Transformational leadership and employee safety performance: a within-person, between-jobs design. *Journal of Occupational Health Psychology* 15 (3): 279–290.

Institute of Occupational Safety and Health (2015). Occupational health risk management in construction: A guide to the key issues of occupational health provision. Wigston, UK: Institution of Occupational Safety and Health. http://www.hse.gov.uk/aboutus/ meetings/iacs/coniac/coniac-oh-guidance.pdf (accessed 13 December 2018).

Institution of Occupational Safety and Health (2015). No time to lose. Respirable crystalline silica: the facts. www.notimetolose.org.uk/wp-content/uploads/2018/03/ Factsheet_Respirable_crystalline_silica_the_facts_MKT2730.pdf (accessed 12 December 2018).

International Atomic Energy Agency (1986). *Summary Report on the Post-accident Review Meeting on the Chernobyl Accident: A Report by the International Nuclear Safety Advisory Group (Safety Series 75-INSAG1)*. Vienna: International Atomic Energy Agency, International Safety Advisory Group.

International Labour Organization (2016). *Workplace Stress: A Collective Challenge*. Geneva: International Labour Organization.

Inyang, N., Al-Hussein, M., El-Rich, M., and Al-Jibouri, S. (2012). Ergonomic analysis and the need for its integration for planning and assessing construction tasks. *Journal of Construction Engineering and Management* 138 (12): 1370–1376.

Iszatt-White, M. (2007). Catching them at it: an ethnography of rule violation. *Ethnography* 8 (4): 445–465.

Jacobson, J. (2011). The ODA's Delivery Partner approach – Creating an integrated framework for mutual success. https://webarchive.nationalarchives.gov. uk/20130403014230/http://learninglegacy.independent.gov.uk/publications/the-oda-s-delivery-partner-approach-creating-an-integrat.php (accessed 27 November 2018).

Jaffar, N., Abdul-Tharim, A.H., Mohd-Kamar, I.F., and Lop, N.S. (2011). A literature review of ergonomics risk factors in construction industry. *Procedia Engineering* 20: 89–97.

Janssen, P.P., Bakker, A.B., and De Jong, A. (2001). A test and refinement of the demand–control–support model in the construction industry. *International Journal of Stress Management* 8 (4): 315–332.

Järvholm, B. (2006). Carcinogens in the construction industry. *Annals of the New York Academy of Sciences* 1076 (1): 421–428.

Jeffcott, S., Pidgeon, N., Weyman, A., and Walls, J. (2006). Risk, trust, and safety culture in U.K. train operating companies. *Risk Analysis: An International Journal* 26 (5): 1105–1121.

Jespersen, A.H. and Hasle, P. (2017). Developing a concept for external audits of psychosocial risks in certified occupational health and safety management systems. *Safety Science* 99: 227–234.

de Jong, A.M. and Vink, P. (2000). The adoption of technological innovations for glaziers; evaluation of a participatory ergonomics approach. *International Journal of Industrial Ergonomics* 26 (1): 39–46.

de Jong, A.M. and Vink, P. (2002). Participatory ergonomics applied in installation work. *Applied Ergonomics* 33 (5): 439–448.

Jorgensen, E., Sokas, R.K., Nickels, L. et al. (2007). An English/Spanish safety climate scale for construction workers. *American Journal of Industrial Medicine* 50: 438–444.

Kadefors, A. (2004). Trust in project relationships – inside the black box. *International Journal of Project Management* 22: 175–182.

Kagioglou, M., Cooper, R., Aouad, G. et al. (1998). *A Generic Guide to the Design and Construction Process Protocol.* Salford: The University of Salford.

Kath, L.M., Marks, K.M., and Ranney, J. (2010). Safety climate dimensions, leader-member exchange, and organisational support as predictors of upward safety communication in a sample of rail industry workers. *Safety Science* 48 (5): 643–650.

Kaukiainen, A., Riala, R., Martikainen, R. et al. (2005). Respiratory symptoms and diseases among construction painters. *International Archives of Occupational and Environmental Health* 78 (6): 452–458.

Kelloway, E.K. and Barling, J. (2010). Leadership development as an intervention in occupational health psychology. *Work & Stress* 24 (3): 260–279.

Kelloway, E.K., Mullen, J., and Francis, L. (2006). Divergent effects of transformational and passive leadership on employee safety. *Journal of Occupational Health Psychology* 11 (1): 76–86.

Kent, D.C. and Becerik-Gerber, B. (2010). Understanding construction industry experience and attitudes toward integrated project delivery. *Journal of Construction Engineering and Management* 136 (8): 815–825.

Kheni, N.A., Dainty, A.R., and Gibb, A. (2008). Health and safety management in developing countries: a study of construction SMEs in Ghana. *Construction Management and Economics* 26 (11): 1159–1169.

Kheni, N.A., Gibb, A.G., and Dainty, A.R. (2010). Health and safety management within small-and medium-sized enterprises (SMEs) in developing countries: study of contextual influences. *Journal of Construction Engineering and Management* 136 (10): 1104–1115.

Kim, S., Nussbaum, M.A., and Jia, B. (2011). Low back injury risks during construction with prefabricated (panelised) walls: effects of task and design factors. *Ergonomics* 54 (1): 60–71.

Kines, P., Lappalainen, J., Mikkelsen, K.L. et al. (2011). Nordic Safety Climate Questionnaire (NOSACQ-50): a new tool for diagnosing occupational safety climate. *International Journal of Industrial Ergonomics* 41 (6): 634–646.

Kinnersley, S. and Roelen, A. (2007). The contribution of design to accidents. *Safety Science* 45: 31–60.

Kittusamy, K. and Buchholz, B. (2004). Whole-body vibration and postural stress among operators of construction equipment: a literature review. *Journal of Safety Research* 35 (3, 2004): 255–261.

Kjellén, U. (2009). The safety measurement problem revisited. *Safety Science* 47: 486–489.

Kleiner, B.M., Hettinger, L.J., DeJoy, D.M. et al. (2015). Sociotechnical attributes of safe and unsafe work systems. *Ergonomics* 58 (4): 635–649.

Knudsen, F. (2009). Paperwork at the service of safety? Workers' reluctance against written procedures exemplified by the concept of 'seamanship'. *Safety Science* 47 (2): 295–303.

Kolmet, M., Marino, R., and Plummer, D. (2006). Anglo-Australian male blue collar workers discuss gender and health issues. *International Journal of Men's Health* 5 (1): 81–91.

Kramer, D.M., Bigelow, P.L., Carlan, N. et al. (2010). Searching for needles in a haystack: identifying innovations to prevent MSDs in the construction sector. *Applied Ergonomics* 41 (4): 577–584.

Kurmis, A. and Apps, S. (2007). Occupationally-acquired noise-induced hearing loss: a senseless workplace hazard. *International Journal of Occupational Medicine and Environmental Health* 20 (2): 127–136.

Lahiri, S., Levenstein, C., Nelson, D.I., and Rosenberg, B.J. (2005). The cost effectiveness of occupational health interventions: prevention of silicosis. *American Journal of Industrial Medicine* 48 (6): 503–514.

LaMontagne, A.D. (2004). Integrating health promotion and health protection in the workplace. In: *Hands-on Health Promotion* (ed. R. Moodie and A. Hulme), 285–298. Sydney: IP Communications.

LaMontagne, A.D., Smith, P.M., Louie, A.M. et al. (2012). Psychosocial and other working conditions: variation by employment arrangement in a sample of working Australians. *American Journal of Industrial Medicine* 55 (2): 93–106.

Lamvik, G.M., Naesje, P.C., Skarholt, K., and Torvatn, H. (2009). Paperwork, management and safety: towards a bureaucratization of working life and a lack of hands-on supervision. In: *Safety, Reliability and Risk Analysis. Theory, Methods and Applications* (ed. S. Martorell, C. Guedes Soares and J. Barnett), 2981–2986. London: Taylor & Francis.

Lancow, J., Ritchie, J., and Crooks, R. (2012). *Infographics: The Power of Visual Storytelling*. Hoboken, NJ: Wiley.

Lang, J., Ochsmann, E., Kraus, T., and Lang, J.W. (2012). Psychosocial work stressors as antecedents of musculoskeletal problems: a systematic review and meta-analysis of stability-adjusted longitudinal studies. *Social Science & Medicine* 75 (7): 1163–1174.

Larsson, S., Poussette, A., and Törner, M. (2008). Psychological climate and safety in the construction industry-mediated influence on safety behaviour. *Safety Science* 46 (3): 405–412.

Latza, U., Pfahlberg, A., and Gefeller, O. (2002). Impact of repetitive manual materials handling and psychosocial work factors on the future prevalence of chronic low-back pain among construction workers. *Scandinavian Journal of Work, Environment & Health* 28: 314–323.

Lawton, R. (1998). Not working to rule: understanding procedural violations at work. *Safety Science* 28: 77–95.

Lazard, A. and Atkinson, L. (2014). Putting environmental infographics center stage: the role of visuals at the elaboration likelihood model's critical point of persuasion. *Science Communication* 37: 6–33.

Lazzarini, R., Duarte, I.A.G., Sumita, J.M., and Minnicelli, R. (2012). Allergic contact dermatitis among construction workers detected in a clinic that did not specialize in occupational dermatitis. *Anais brasileiros de dermatologia* 87 (4): 567–571.

Leah, C., Riley, D., and Jones, A. (2013). *Mobile Elevated Work Platform (MEWP) Incident Analysis*. London: HSE Books.

Lee, J., Mahendra, S., and Alvarez, P.J. (2010). Nanomaterials in the construction industry: a review of their applications and environmental health and safety considerations. *ACS Nano* 4 (7): 3580–3590.

Leensen, M.C.J., Van Duivenbooden, J.C., and Dreschler, W.A. (2011). A retrospective analysis of noise-induced hearing loss in the Dutch construction industry. *International Archives of Occupational and Environmental Health* 84 (5): 577–590.

Leka, S., Van Wassenhove, W., and Jain, A. (2015). Is psychosocial risk prevention possible? Deconstructing common presumptions. *Safety Science* 71: 61–67.

LePlat, J. (1998). About implementation of safety rules. *Safety Science* 29: 189–204.

Lester, P.M. (2011). *Visual Communication: Images with Messages*, 5e. Independence, KY/ Wadsworth: Cengage Learning.

Leung, M.Y., Chan, Y.S., and Olomolaiye, P. (2008). Impact of stress on the performance of construction project managers. *Journal of Construction Engineering and Management* 134 (8): 644–652.

Levitt, R.E. and Samelson, N.M. (1993). *Construction Safety Management*, 2e. New York: Wiley.

Li, Y. and Guldenmund, F.W. (2018). Safety management systems: a broad overview of the literature. *Safety Science* 103: 94–123.

Lingard, H. and Francis, V. (2004). The work-life experiences of office and site-based employees in the Australian construction industry. *Construction Management and Economics* 22 (9): 991–1002.

Lingard, H. and Francis, V. (2005a). Does work–family conflict mediate the relationship between job schedule demands and burnout in male construction professionals and managers? *Construction Management and Economics* 23 (7): 733–745.

Lingard, H. and Francis, V. (2005b). The decline of the "traditional" family: work-life benefits as a means of promoting a diverse workforce. *Construction Management and Economics* 23: 1045–1057.

Lingard, H. and Francis, V. (2006). Does a supportive work environment moderate the relationship between work-family conflict and burnout among construction professionals? *Construction Management and Economics* 24 (2): 185–196.

Lingard, H. and Francis, V. (2007). "Negative interference" between Australian construction professionals' work and family roles: evidence of an asymmetrical relationship. *Engineering, Construction and Architectural Management* 14 (1): 79–93.

Lingard, H. and Francis, V. (2009). *Managing Work-Life Balance in Construction*. Routledge.

Lingard, H. and Holmes, N. (2001). Understandings of occupational health and safety risk control in small business construction firms: barriers to implementing technological controls. *Construction Management and Economics* 19 (2): 217–226.

Lingard, H. and Rowlinson, S.M. (2005). *Occupational Health and Safety in Construction Project Management*. UK: Taylor & Francis.

Lingard, H. and Turner, M. (2015). Improving the health of male, blue collar construction workers: a social ecological perspective. *Construction Management and Economics* 33 (1): 18–34.

Lingard, H. and Turner, M. (2017). Promoting construction workers' health: a multi-level system perspective. *Construction Management and Economics* 35 (5): 239–253.

Lingard, H., Tombesi, P., Blismas, N., and Gardiner, B. (2007). Guilty in theory or responsible in practice? Architects and the decisions affecting occupational health and safety in construction design. In: *Looking Ahead: Defining the Terms of a Sustainable Architectural Profession* (ed. P. Tombesi, B. Gardiner and T. Mussen), 49–59. Manuka ACT: Royal Australian Institute of Architects.

Lingard, H., Blismas, N., Cooke, T., and Cooper, H. (2009a). The model client framework: resources to help Australian government agencies to promote safe construction. *International Journal of Managing Projects in Business* 2 (1): 131–140.

Lingard, H., Cooke, T., and Blismas, N. (2009b). Group-level safety climate in the Australian construction industry: within-group homogeneity and between-group differences in road construction and maintenance. *Construction Management and Economics* 27 (4): 419–432.

Lingard, H., Francis, V., and Turner, M. (2010a). Work-family conflict in construction: case for finer-grained analysis. *Journal of Construction Engineering and Management* 136 (11): 1196–1206.

Lingard, H., Cooke, T., and Blismas, N. (2010b). Safety climate in conditions of construction subcontracting: a multi-level analysis. *Construction Management and Economics* 28: 813–825.

Lingard, H., Cooke, T., and Blismas, N. (2010c). Properties of group safety climate in construction: the development and evaluation of a typology. *Construction Management and Economics* 28: 1099–1112.

Lingard, H., Wakefield, R., and Cashin, P. (2011a). The development and testing of a hierarchical measure of project OHS performance. *Engineering, Construction and Architectural Management* 18: 30–49.

Lingard, H., Cooke, T., and Blismas, N. (2011b). Who is 'the designer' in construction occupational health and safety? In: *ARCOM Twenty-Seventh Annual Conference, Conference Proceedings*, 299–308. Bristol, 5–7 September,.

Lingard, H., Cooke, T., and Blismas, N. (2012a). Designing for construction workers' occupational health and safety: a case study of socio-material complexity. *Construction Management and Economics* 30 (5): 367–382.

Lingard, H., Cooke, T., and Blismas, N. (2012b). Do perceptions of supervisors' safety responses mediate the relationship between perceptions of the organizational safety climate and incident rates in construction supply chain? *Journal of Construction Engineering and Management* 138 (2): 234–241.

Lingard, H., Francis, V., and Turner, M. (2012c). Work time demands, work time control and supervisor support in the Australian construction industry: an analysis of work-family interaction. *Engineering, Construction and Architectural Management* 19 (6): 647–665.

Lingard, H., Wakefield, R., and Blismas, N. (2013a). "If you cannot measure it, you cannot improve it": measuring health and safety performance in the construction industry. In: *Proceedings of the 19th Triennial CIB World Building Congress* (ed. J.V. McCarthy AO, S. Kajewski, K. Manley and K. Hampson), 1–2. Brisbane, Queensland: Queensland University of Technology.

Lingard, H., Cooke, T., Blismas, N., and Wakefield, R. (2013b). Prevention through design: trade-offs in reducing occupational health and safety risk for the construction and operation of a facility. *Built Environment Project and Asset Management* 3 (1): 7–23.

Lingard, H., Pirzadeh, P., Blismas, N. et al. (2014). Exploring the link between early constructor involvement in project decision-making and the efficacy of health and safety risk control. *Construction Management and Economics* 32 (9): 918–931.

Lingard, H., Pirzadeh, P., Blismas, N. et al. (2015a). The relationship between pre-construction decision-making and the quality of risk control: testing the time-safety influence curve. *Engineering, Construction and Architectural Management* 22 (1): 108–124.

Lingard, H., Pink, S., Harley, J., and Edirisinghe, R. (2015b). Looking and learning: using participatory video to improve health and safety in the construction industry. *Construction Management and Economics* 33 (9): 741–752.

Lingard, H., Blismas, N., Zhang, R.P., et al. (2015c). Engaging stakeholders in improving the quality of OHS decision-making in construction projects: Research to Practice Report. RMIT University, Melbourne: Centre for Construction Work Health and Safety Research https://designforconstructionsafety.files.wordpress.com/2018/05/research-to-practice-report_final-2015.pdf (accessed 13 December 2018).

Lingard, H., Peihua Zhang, R., Blismas, N. et al. (2015d). Are we on the same page? Exploring construction professionals' mental models of occupational health and safety. *Construction Management and Economics* 33 (1): 73–84.

Lingard, H., Edirisinghe, R., and Harley, J. (2015d). *Using Digital Technology to Share Health and Safety Knowledge.* Melbourne: Centre for Construction Work Health and Safety Research, RMIT University.

Lingard, H., Pink, S., Hayes, J. et al. (2016). Using participatory video to understand subcontracted construction workers' safety rule violations. In: *Proceedings of the 32nd Annual ARCOM Conference*, 5–7 September 2016, vol. 1 (ed. P.W. Chan and C.J. Neilson), 497–506. Manchester, UK: Association of Researchers in Construction Management.

Lingard, H., Hallowell, M., Salas, R., and Pirzadeh, P. (2017a). Leading or lagging? Temporal analysis of safety indicators on a large infrastructure construction project. *Safety Science* 91: 206–220.

Lingard, H., Harley, J., Zhang, R., and Ryan, G. (2017b). Work Health and Safety Culture in the ACT Construction Industry. RMIT University, Melbourne: Centre for Construction Work Health and Safety Research. https://www.cmtedd.act.gov.au/__data/assets/pdf_file/0020/1121744/Work-Health-Safety-Culture-ACT-Construction-Industry.pdf (accessed 13 December 2018).

Lingard, H., Blismas, N., Harley, J. et al. (2018). Making the invisible visible: stimulating work health and safety-relevant thinking through the use of infographics in construction design. *Engineering, Construction and Architectural Management* 25 (1): 39–61.

Lofquist, E.A. (2010). The art of measuring nothing: the paradox of measuring safety in a changing civil aviation industry using traditional safety metrics. *Safety Science* 48: 1520–1529.

Löfstedt, R.E. (2011). *Reclaiming Health and Safety for all: An Independent Review of Health and Safety Legislation.* London: Department for Work and Pensions.

Lombardi, D.A., Verma, S.K., Brennan, M.J., and Perry, M.J. (2009). Factors influencing worker use of personal protective eyewear. *Accident Analysis and Prevention* 41 (4): 755–762.

Loosemore, M. and Andonakis, N. (2007). Barriers to implementing OHS reforms – the experiences of small subcontractors in the Australian Construction Industry. *International Journal of Project Management* 25 (6): 579–588.

Loosemore, M. and Galea, N. (2008). Genderlect and conflict in the Australian construction industry. *Construction Management and Economics* 26 (2): 125–135.

Love, P., Gunasekaran, A., and Li, H. (1998). Concurrent engineering: a strategy for procuring construction projects. *International Journal of Project Management* 16 (6): 375–383.

Love, P.E.D., Edwards, D.J., and Irani, Z. (2010). Work stress, support, and mental health in construction. *Journal of Construction Engineering and Management* 136 (6): 650–658.

Lu, C.S. and Shang, K.C. (2005). An empirical investigation of safety climate in container terminal operators. *Journal of Safety Research* 36: 297–308.

Ludewig, P.M. and Borstad, J.D. (2003). Effects of a home exercise programme on shoulder pain and functional status in construction workers. *Occupational and Environmental Medicine* 60 (11): 841–849.

Luijsterburg, P.A., Bongers, P.M., and de Vroome, E.M. (2005). A new bricklayers' method for use in the construction industry. *Scandinavian Journal of Work, Environment & Health* 31: 394–400.

Lundberg, J., Rollenhagen, C., and Hollnagel, E. (2009). What-you-look-for-is-what you-find: the consequences of underlying accident models in eight accident investigation manuals. *Safety Science* 47: 1297–1311.

Lynch, J.W., Kaplan, G.A., and Salonen, J.T. (1997). Why do poor people behave poorly? Variation in adult health behaviours and psychosocial characteristics by stages of the socioeconomic lifecourse. *Social Science & Medicine* 44 (6): 809–819.

MacKenzie, S. (2008). *A Close Look at Work and Life Balance/Wellbeing in the Victorian Commercial Building and Construction Sector*. Melbourne: Building Industry Consultative Council.

Maiden, S. (2010). Foil insulation program suspended after deaths. *The Australian*, February 9th, 2010, www.theaustralian.com.au/national-affairs/climate/foil-insulation-program-suspeended-aafter-deaths/story-e6frg6xf-1225828406965 (accessed 13 December 2018).

Manu, P., Ankrah, N., Proverbs, D., and Suresh, S. (2010). An approach for determining the extent of contribution of construction project features to accident causation. *Safety Science* 48: 687–692.

Manu, P., Ankrah, N., Proverbs, D., and Suresh, S. (2013). Mitigating the health and safety influence of subcontracting in construction: the approach of main contractors. *International Journal of Project Management* 31 (7): 1017–1026.

Manuele, F.A. (2006). Achieving risk reduction, effectively. *Process Safety and Environmental Protection* 84 (3): 184–190.

Marchant, G.E., Sylvester, D.J., and Abbott, K.W. (2008). Risk management principles for nanotechnology. *NanoEthics* 2 (1): 43–60.

Mason, S., Lawton, B., Travers, V., Rycraft, H., Ackroyd, P. and Collier, S., (1995), Improvingcompliance with safety procedures. *Sudbury*, UK: HSE Books.

Mayhew, C. and Quinlan, M. (1997). Subcontracting and occupational health and safety in the residential building industry. *Industrial Relations Journal* 28 (3): 192–205.

Mayhew, C., Quinlan, M., and Ferris, R. (1997). The effects of subcontracting/outsourcing on occupational health and safety: survey evidence from four Australian industries. *Safety Science* 25: 163–178.

McAllister, M. and McKinnon, J. (2009). The importance of teaching and learning resilience in the health disciplines: a critical review of the literature. *Nurse Education Today* 29 (4): 371–379.

McDonald, G., Jackson, D., Wilkes, L., and Vickers, M.H. (2013). Personal resilience in nurses and midwives: effects of a work-based educational intervention. *Contemporary Nurse* 45 (1): 134–143.

McIntosh, W.L., Spies, E., Stone, D.M. et al. (2016). Suicide rates by occupational group — 17 states, 2012. *Morbidity and Mortality Weekly Report, Centers for Disease Control and Prevention* 65: 641–645.

McLaren, L. and Hawe, P. (2005). Ecological perspectives in health research. *Journal of Epidemiology & Community Health* 59 (1): 6–14.

McLeroy, K.R., Bibeau, D., Steckler, A., and Glanz, K. (1988). An ecological perspective on health promotion programs. *Health Education & Behavior* 15 (4): 351–377.

Mearns, K. (2009). From reactive to proactive – can LPIs deliver? *Safety Science* 47 (4): 491–492.

Mearns, K. and Flin, R. (1999). Assessing the state of organisational safety — culture or climate? *Current Psychology* 18 (1): 5–17.

Mearns, K. and Yule, S. (2009). The role of national culture in determining safety performance: challenges for the global oil and gas industry. *Safety Science* 47 (6): 777–785.

Mearns, K., Whitaker, S.M., and Flin, R. (2003). Safety climate, safety management practice and safety performance in offshore environments. *Safety Science* 419 (8): 641–680.

Meerding, W.J., IJzelenberg, W., Koopmanschap, M.A. et al. (2005). Health problems lead to considerable productivity loss at work among workers with high physical load jobs. *Journal of Clinical Epidemiology* 58 (5): 517–523.

Meldrum, A., Hare, B., and Cameron, I. (2009). Road testing a health and safety worker engagement tool-kit in the construction industry. *Engineering, Construction and Architectural Management* 16 (6): 612–632.

Melia, J.L., Mearns, K., Silva, S.A., and Lima, M.L. (2008). Safety climate responses and the perceived risk of accidents in the construction industry. *Safety Science* 46: 949–958.

Mengolinim, A. and Debarberis, L. (2008). Effectiveness evaluation methodology for safety processes to enhance organizational safety culture in hazardous installations. *Journal of Hazardous Materials* 155: 243–252.

Merlino, L.A., Rosecrance, J.C., Anton, D., and Cook, T.M. (2003). Symptoms of musculoskeletal disorders among apprentice construction workers. *Applied Occupational and Environmental Hygiene* 18 (1): 57–64.

Milner, A., Page, K., Spencer-Thomas, S., and Lamotagne, A.D. (2015). Workplace suicide prevention: a systematic review of published and unpublished activities. *Health Promotion International* 30 (1): 29–37.

Mirka, G.A., Monroe, M., Nay, T. et al. (2003). Ergonomic interventions for the reduction of low back stress in framing carpenters in the home building industry. *International Journal of Industrial Ergonomics* 31 (6): 397–409.

Mitropoulos, P. and Cupido, G. (2009). Safety as an emergent property: investigation into the work practices of high-reliability framing crews. *Journal of Construction Engineering and Management* 135 (5): 407–415.

Mohaghegh, Z. and Mosleh, A. (2009). Measurement techniques for organizational safety causal models: characterization and suggestions for enhancements. *Safety Science* 47: 1398–1409.

Moir, S. and Buchholz, B. (1996). Emerging participatory approaches to ergonomic interventions in the construction industry. *American Journal of Industrial Medicine* 29 (4): 425–430.

Mullen, J.E. (2005). Testing a model of employee willingness to raise safety issues. *Canadian Journal of Behavioral Sciences* 37 (4): 259–268.

Mullen, J.E. and Kelloway, E.K. (2009). Safety leadership: a longitudinal study of the effects of transformational leadership on safety outcomes. *Journal of Occupational and Organisational Psychology* 82 (2): 253–272.

Mullen, J.E., Kelloway, E.K., and Teed, M. (2011). Inconsistent style of leadership as a predictor of safety behaviour. *Work & Stress* 25 (1): 41–54.

Murphy, C. (2010). The Planning and Management of Perth Arena: Media Statement by the Auditor General for Western Australia. https://audit.wa.gov.au/wpcontent/uploads/2013/05/report2010_01.pdf (accessed 13 December 2018).

den Otter, A. and Emmitt, S. (2008). Design team communication and design task complexity: the preference for dialogues. *Architectural, Engineering and Design Management* 4 (2): 121–129.

Naevestad, T.O. (2009). Mapping research on culture and safety in high-risk organisations: arguments for a sociotechnical understanding of safety culture. *Journal of Contingencies and Crisis Management* 7 (2): 126–136.

Nahrgang, J.D., Morgeson, F.P., and Hofmann, D.A. (2011). Safety at work: a meta-analytic investigation of the link between job demands, job resources, burnout, engagement, and safety outcomes. *Journal of Applied Psychology* 96 (1): 71.

National Institute for Occupational Health and Safety (2008). *National Occupational Research Agenda, National Construction Agenda*. Atlanta, GA: Centre for Disease Control/NIOSH.

National Occupational Safety and Health Commission (1999). *OHS Performance Measurement in the Construction Industry: Development of Positive Performance Indicators*. Canberra: AGPS.

Neal, A. and Griffin, M.A. (2002). Safety climate and safety behaviour. *Australian Journal of Management* 27 (1_suppl): 67–75.

Neal, A. and Griffin, M.A. (2006). A study of the lagged relationships among safety climate, safety motivation, group safety climate, safety behaviour and accidents at individual and group levels. *Journal of Applied Psychology* 91: 946–953.

Neitzel, R. and Seixas, N. (2005). The effectiveness of hearing protection among construction workers. *Journal of Occupational and Environmental Hygiene* 2 (4): 227–238.

Nicolini, D., Holti, R., and Smalley, M. (2001). Integrating project activities: the theory and practice of managing the supply chain through clusters. *Construction Management and Economics* 19: 37–47.

Nielsen, K. (2000). Organizational theories implicit in various approaches to OHS management. In: *Systematic Occupational Health and Safety Management: Perspectives on an International Development* (ed. K. Frick, P.L. Jensen, M. Quinlan and T. Wilthagen), 99–124. Amsterdam: Pergamon Press.

Nielsen, K.J., Hansen, C.D., Bloksgaard, L. et al. (2015). The impact of masculinity on safety oversights, safety priority and safety violations in two male-dominated occupations. *Safety Science* 76: 82–89.

Niu, M., Leicht, R. M., and Rowlinson, S. (2015). Exploring the Role of Owner Construction Site. Working Paper Proceedings of the Engineering Project Organization Conference (EPOC), The University of Edinburgh, Scotland, UK, June 24–26.

Noblet, A. and LaMontagne, A.D. (2006). The role of workplace health promotion in addressing job stress. *Health Promotion International* 21 (4): 346–353.

Noone, P. (2013). Keeping baby boomer construction workers working. *Occupational Medicine* 63 (3): 244–245.

Nordlöf, H., Wiitavaara, B., Winblad, U. et al. (2015). Safety culture and reasons for risk-taking at a large steel manufacturing company: investigating the worker perspective. *Safety Science* 73: 126–135.

Occupational Health and Safety Act 2004 (Vic.) (2004). http://www.legislation.vic.gov.au/Domino/Web_Notes/LDMS/PubLawToday.nsf (accessed 27 November 2018).

Occupational Health and Safety Administration (2018). Silica. https://www.osha.gov/dsg/topics/silicacrystalline/gi_maritime.html (accessed 20 March 2018).

Occupational Safety and Health Administration (2002). Crystalline Silica Exposure: Health Hazard Information. https://www.osha.gov/OshDoc/data_General_Facts/crystalline-factsheet.pdf (accessed 13 December 2018).

O'Dea, A. and Flin, R. (2001). Site managers and safety leadership in the offshore oil and gas industry. *Safety Science* 37 (1): 39–57.

Office of the Federal Safety Commissioner (2007). The Model Client – Promoting Safe Construction. www.fsc.gov.au/sites/fsc/resources/az/pages/themodelclient-promotingsafeconstruction (accessed 27 November 2018).

Olive, C., O'Connor, T.M., and Mannan, M.S. (2006). Relationship of safety culture and process safety. *Journal of Hazardous Materials* 130 (1–2): 133–140.

Oquendo, M.A., Perez-Rodriguez, M.M., Poh, E. et al. (2014). Life events: a complex role in the timing of suicidal behavior among depressed patients. *Molecular Psychiatry* 19 (8): 902–909.

Oswald, D. (2016). Investigating unsafe acts on a large multinational construction project. PhD Thesis, University of Edinburgh. UK.

Oswald, D., Sherratt, F., and Smith, S. (2013). Exploring factors affecting unsafe behaviours in construction. In: *Proceedings of 29th Annual ARCOM Conference*, 2–4 September 2013, (ed. S.D. Smith and D.D. Ahiaga-Dagbui), 335–344. Reading, UK: Association of Researchers in Construction Management.

Otten, J.J., Cheng, K., and Drewnowski, A. (2015). Infographics and public policy: using data visualization to convey complex information. *Health Affairs* 34 (11): 1901–1907.

Oude Hengel, K.M., Joling, C.I., Proper, K.I. et al. (2010). A worksite prevention program for construction workers: design of a randomized controlled trial. *BMC Public Health* 10: 336. http://www.biomedcentral.com/1471-2458/10/336 (accessed 13 December 2018).

Oude Hengel, K.M., Blatter, B.M., Geuskens, G.A. et al. (2011). Factors associated with the ability and willingness to continue working until the age of 65 in construction workers. *International Archives of Occupational and Environmental Health* https://doi.org/10.1007/s00420-011-0719-3.

Oude Hengel, K.M., Blatter, B.M., Joling, C.I. et al. (2012). Effectiveness of an intervention at construction worksites on work engagement, social support, physical workload and the need for recovery: results from a cluster randomized controlled trial. *BMC Public Health* 12: https://doi.org/10.1186/1471-2458-12-1008.

Oude Hengel, K.M., Blatter, B., van der Molen, H.F. et al. (2013). The effectiveness of a construction worksite prevention program on work ability, health and sick leave: results from a cluster randomized controlled trial. *Scandinavian Journal of Work, Environment & Health* 39 (5): 456–466.

Ozmec, M.N., Larlsen, I.L., Kines, P. et al. (2015). Negotiating safety practice in small construction companies. *Safety Science* 71: 275–281.

Parida, R. and Ray, P.K. (2012). Study and analysis of occupational risk factors for ergonomic design of construction worksystems. *Work* 41: 3788–3794.

Parker, M. (2000). *Organisational Culture and Identity*. London: Sage Publications.

Parker, S.K., Axtell, C.M., and Turner, N. (2001). Designing a safer workplace: importance of job autonomy, communication quality, and supportive supervisors. *Journal of Occupational Health Psychology* 6 (3): 211–228.

Parker, D., Lawrie, M., and Hudson, P. (2006). A framework for understanding the development of organisational safety culture. *Safety Science* 44 (6): 551–562.

Pauwels, L. (2000). Taking the visual turn in research and scholarly communication: key issues in developing a more visually literate (social) science. *Visual Sociology* 15 (1): 7–14.

Pedersen, L.M., Nielsen, K.J., and Kines, P. (2012). Realistic evaluation as a new way to design and evaluate occupational safety interventions. *Safety Science* 50 (1): 48–54.

Peterson, J.S. and Zwerling, C. (1998). Comparison of health outcomes among older construction and blue-collar employees in the United States. *American Journal of Industrial Medicine* 34 (3): 280–287.

Pidgeon, N. (1991). Safety culture and risk management in organisations. *Journal of Cross-Cultural Psychology* 22 (1): 129–140.

Pidgeon, N. (1998). Risk assessment, risk values and the social science programme: why we do need risk perception research. *Reliability Engineering & System Safety* 59 (1): 5–15.

Pietroforte, R. (1995). Cladding systems: technological change and design arrangements. *Journal of Architectural Engineering* 1 (3): 100–107.

Pietroforte, R. (1997). Communication and governance in the building process. *Construction Management and Economics* 15: 71–82.

Pilbeam, C., Davidson, R., Doherty, N., and Denyer, D. (2016). What learning happens? Using audio diaries to capture learning in response to safety-related events within retain and logistics organizations. *Safety Science* 81: 59–67.

Polanyi, M. (1958). *Personal Knowledge: Towards a Post-Critical Philosophy*. Chicago, IL: University of Chicago Press.

Portewig, T.C. (2004). Making sense of the visual in technical communication: a visual literacy approach to pedagogy. *Journal of Technical Writing and Communication* 34 (1): 31–42.

Potter, R.E., Dollard, M.F., Owen, M.S. et al. (2017). Assessing a national work health and safety policy intervention using the psychosocial safety climate framework. *Safety Science* 100: 91–102.

Pousette, A., Larsson, S., and Törner, M. (2008). Safety climate cross-validation, strength and prediction of safety behaviour. *Safety Science* 46: 398–404.

PriceWaterhouseCoopers. (2014). Creating a mentally healthy workplace: Return on investment analysis. www.headsup.org.au/docs/default-source/resources/bl1269-brochure---pwc-roi-analysis.pdf?sfvrsn=6 (accessed 13 December 2018).

Priemus, H. and Ale, B. (2010). Construction safety: an analysis of systems failure: the case of the multifunctional Bos & Lommerplein estate, Amsterdam. *Safety Science* 48 (2): 111–122.

Pritchard, C. and McCarthy, A. (2002). Promoting health in the construction industry? *Journal of Occupational and Environmental Medicine* 44 (6): 540–545.

Priyadarshani, K., Karunasena, G., and Jayasuriya, S. (2013). Construction safety assessment framework for developing countries: a case study of Sri Lanka. *Journal of Construction in Developing Countries* 18 (1): 33–51.

Probst, T.M., Barbaranelli, C., and Petitta, L. (2013). The relationship between job insecurity and accident under-reporting: a test in two countries. *Work & Stress* 27 (4): 383–402.

Pronk, N.P. (2013). Integrated worker health protection and promotion programs: overview and perspectives on health and economic outcomes. *Journal of Occupational and Environmental Medicine* 55 (12 Suppl): S30–S37.

Prussia, G.E., Brown, K.A., and Willis, P.G. (2003). Mental models of safety: do managers and employees see eye to eye? *Journal of Safety Research* 34 (2): 143–156.

Quinlan, M. (2011). *Supply Chains and Networks*. Canberra: Safe Work Australia.

Railnews (2012). Edinburgh tram costs soar again. www.railnews.co.uk/news/2012/06/14-edinburgh-tram-costssoar-again.html (accessed 13 December 2018).

Rasmussen, J. (1982). Human errors. A taxonomy for describing human malfunction in industrial installations. *Journal of Occupational Accidents* 4 (2–4): 311–333.

Reason, J. (1990). *Human Error*. Cambridge: Cambridge University Press.

Reason, J. (1997). *Managing the Risks of Organisational Accidents*. Aldershot: Ashgate.

Reason, J. (1998). Achieving a safe culture: theory and practice. *Work & Stress* 12 (3): 293–306.

Reason, J. (2000). Safety paradoxes and safety culture. *Injury Control and Safety Promotion* 7 (1): 3–14.

Reason, J. (2008). *The Human Contribution: Unsafe Acts, Accidents and Heroic Recoveries*. Surrey: Ashgate Publishing Company.

Reason, J. (2013). *A Life in Error*. Farnham, UK: Ashgate.

Redinger, C.F. and Levine, S.P. (1998). Development and evaluation of the Michigan Occupational Health and Safety Management System assessment instrument: a universal OHSMS performance measurement tool. *American Industrial Hygiene Association* 59 (8): 572–581.

Reiman, T. and Pietkainen, E. (2012). Leading indicators of system safety – monitoring and driving the organizational safety potential. *Safety Science* 50: 1993–2000.

Reiman, T. and Rollenhagen, C. (2011). Human and organizational biases affecting the management of safety. *Reliability Engineering & System Safety* 96 (10): 1263–1274.

Rempel, D. and Barr, A. (2015). A universal rig for supporting large hammer drills: Reduced injury risk and improved productivity. *Safety science* 78: 20–24.

Rempel, D., Star, D., Barr, A. et al. (2009). A new method for overhead drilling. *Ergonomics* 52 (12): 1584–1589.

Rempel, D., Star, D., Barr, A. et al. (2010a). Field evaluation of a modified intervention for overhead drilling. *Journal of Occupational and Environmental Hygiene* 7 (4): 194–202.

Rempel, D., Star, D., Barr, A., and Janowitz, I. (2010b). Overhead drilling: comparing three bases for aligning a drilling jig to vertical. *Journal of Safety Research* 41 (3): 247–251.

Richter, A. and Koch, C. (2004). Integration, differentiation and ambiguity in safety cultures. *Safety Science* 42 (8): 703–722.

Rinder, M.M., Genaidy, A., Salem, S. et al. (2008). Interventions in the construction industry: a systematic review and critical appraisal. *Human Factors and Ergonomics in Manufacturing & Service Industries* 18 (2): 212–229.

Ringen, K. and Englund, A. (2006). The construction industry. *Annals of the New York Academy of Sciences* 1076 (1): 388–393.

Ringen, K., Dement, J., Welch, L. et al. (2014). Risks of a lifetime in construction. Part II: Chronic occupational diseases. *American Journal of Industrial Medicine* 57 (11): 1235–1245.

Rivilis, I., Cole, D.C., Frazer, M.B. et al. (2006). Evaluation of a participatory ergonomic intervention aimed at improving musculoskeletal health. *American Journal of Industrial Medicine* 49 (10): 801–810.

Rivilis, I., Van Eerd, D., Cullen, K. et al. (2008). Effectiveness of participatory ergonomic interventions on health outcomes: a systematic review. *Applied Ergonomics* 39 (3): 342–358.

Roberts, S.E., Jaremin, B., and Lloyd, K. (2013). High-risk occupations for suicide. *Psychological Medicine* 43 (6): 1231–1240.

Robertson, D. and Fox, J. (2000). Industrial use of safety-related expert systems (Contract Research Report 296/2000). Norwich, UK: Her Majesty's Stationery Office. http://www.hse.gov.uk/research/crr_pdf/2000/crr00296.pdf (accessed 13 December 2018).

Robson, L.S., Shannon, H.S., Goldenhar, L. M., and Hale, A.R. (2001). Guide to Evaluating the Effectiveness of Strategies for Preventing Work Injuries: How to show whether a safety intervention really works. Cincinnati, OH: National Institute for Occupational Safety and Health. https://www.cdc.gov/niosh/docs/2001-119/pdfs/2001-119.pdf (accessed 13 December 2018).

Rollenhagen, C. (2010). Can focus on safety culture become an excuse for not rethinking design of technology? *Safety Science* 48 (2): 268–278.

Roos, E., Lahelma, E., and Rahkonen, O. (2006). Work–family conflicts and drinking behaviours among employed women and men. *Drug and Alcohol Dependence* 83 (1): 49–56.

Rouse, W.B. and Morris, N.M. (1986). On looking into the black box: prospects and limits in the search for mental models. *Psychological Bulletin* 100 (3): 349–363.

Routley, V.H. and Ozanne-Smith, J.E. (2012). Work-related suicide in Victoria, Australia: a broad perspective. *International Journal of Injury Control and Safety Promotion* 19 (2): 131–134.

Ryff, C.D. and Singer, B. (2003). Flourishing under fire: resilience as a prototype of challenged thriving. In: *Flourishing: Positive Psychology and the Life Well-Lived* (ed. C.L.M. Keyes and J. Haidt), 15–36. Washington: American Psychological Association.

Safe Work Australia (2012). *Australian Work Health and Safety Strategy 2012–2022*. Canberra: Safe Work Australia www.safeworkaustralia.gov.au/system/files/documents/1804/australian-work-health-safety-strategy-2012-2022v2_1.pdf (accessed 12 December 2018).

Safe Work Australia (2015a). *Construction Industry Profile*. Canberra: Australian Government.

Safe Work Australia (2015b). *Exposure to Multiple Hazards among Australian Workers*. Canberra: Safe Work Australia.

Safe Work Australia (2017). Construction Industry Profile. https://www.safeworkaustralia.gov.au/system/files/documents/1702/construction-industry-profile.pdf (accessed 13 December 2018).

Safe Work Australia (2018), Priority industry snapshot: Construction. https://www.safeworkaustralia.gov.au/system/files/documents/1807/construction-priority-industry-snapshot-2018.pdf (accessed 13 December 2018).

Safe Work Australia (2018a). Crystalline silica and silicosis. www.safeworkaustralia.gov.au/silica (accessed 11 December 2018).

Safe Work Australia (2018b). Workplace exposure standards for airborne contaminants. www.safeworkaustralia.gov.au/exposure-standards (accessed 20 June 2018).

Saksvik, P.Ø. and Quinlan, M. (2003). Regulating systematic occupational health and safety management: comparing the Norwegian and Australian experience. *Relations Industrielles* 58 (1): 33–59.

Salas, R. and Hallowell, M. (2016). Predictive validity of safety leading indicators: empirical assessment in the oil and gas sector. *Journal of Construction Engineering and Management* 04016052:1–11. https://doi.org/10.1061/(ASCE)CO.1943-7862.0001167.

Salas, E., Sims, D.E., and Burke, C.S. (2005). Is there a "big five" in teamwork? *Small Group Research* 36 (5): 555–599.

Sallis, J.F., Owen, N., and Fisher, E.B. (2008). Ecological models of health behavior. In: *Health Behavior and Health Education: Theory, Research, and Practice*, 4e (ed. K. Glanz, B.K. Rimer and K. Viswanath), 465–486. San Francisco: Jossey-Bass.

Schein, E.H. (1999). *The Corporate Culture Survival Guide*. San Francisco: Jossey-Bass.

Schein, E.H. (2006). *Organisational Culture and Leadership*, 3e. San Francisco: Jossey-Bass.

Schein, E.H. (2010). *Organisational Culture and Leadership*, 4e. San Francisco: Jossey-Bass.

Schill, A.L. and Chosewood, L.C. (2013). The NIOSH total worker health™ program: an overview. *Journal of Occupational and Environmental Medicine* 55: S8–S11.

Schneider, S. (2001). Musculoskeletal injuries in construction: a review of the literature. *Applied Occupational and Environmental Hygiene* 16 (11): 1056–1064.

Schulte, P.A., Rinehart, R., Okun, A. et al. (2008). National prevention through design (PtD) initiative. *Journal of Safety Research* 39 (2): 115–121.

Schulte, P.A., Pandalai, S., Wulsin, V., and Chun, H. (2012). Interaction of occupational and personal risk factors in workforce health and safety. *American Journal of Public Health* 102 (3): 434–448.

Schwartz-Lifshitz, M., Zalsman, G., Giner, L., and Oquendo, M.A. (2012). Can we really prevent suicide? *Current Psychiatry Reports* 14 (6): 624–633.

Seo, D.C. (2005). An explicative model of unsafe work behavior. *Safety Science* 43 (3): 187–211.

Shaw, J. and Robertson, C. (1997). *Participatory Video: A Practical Guide to Using Video Creatively in Group Development Work*. London, UK: Routledge.

Shea, T., De Cieri, H., Donohue, R. et al. (2016). Leading indicators of occupational health and safety: an employee and workplace level validation study. *Safety Science* 85: 293–304.

Sherratt, F. (2014). Exploring 'Zero Target' safety programmes in the UK construction industry. *Construction Management and Economics* 32 (7-8): 737–748.

Sherratt, F. (2015). Legitimizing public health control on sites? A critical discourse analysis of the Responsibility Deal Construction Pledge. *Construction Management and Economics* 33 (5-6): 444–452.

Sherratt, F. (2016). *Unpacking Construction Site Safety*. UK: Wiley.

Sherratt, F., Farrell, P., and Noble, R. (2013). UK construction site safety: discourses of enforcement and engagement. *Construction Management and Economics* 31 (6): 623–635.

Shirom, A., Melamed, S., Toker, S. et al. (2005). Burnout and health review: current knowledge and future research directions. In: *International Review of Industrial and Organizational Psychology*, vol. 20 (ed. G.P. Hodgkinson and K. Ford), 269–309. Palo Alto, CA: Consulting Psychologists Press.

Siebert, U., Rothenbacher, D., Daniel, U., and Brenner, H. (2001). Demonstration of the healthy worker survivor effect in a cohort of workers in the construction industry. *Occupational and Environmental Medicine* 58 (9): 774–779.

Simoms, T. (2002). Behavioral integrity: the perceived alignment between managers' words and deeds as a research focus. *Organisation Science* 13 (1): 18–35.

Sinelnikov, S., Inouye, J., and Kerper, S. (2015). Using leading indicators to measure occupational health and safety performance. *Safety Science* 72: 240–248.

Siu, O., Phillips, D.R., and Leung, T. (2004). Safety climate and safety performance among construction workers in Hong Kong: the role of psychological strains as mediators. *Accident Analysis and Prevention* 36: 359–366.

Slaughter, S. (1993). Buildings as sources of construction innovation. *Journal of Construction Engineering and Management* 119 (3): 532–549.

Smallwood, J. (2012). Mass of materials: the impact of designers on construction ergonomics. *Work* 41: 5425–5430.

Smallwood, J.J., Haupt, T.C., and Musonda, I. (2009). Client attitude to health and safety – a report on contractor's perceptions. *Acta Structilia* 16 (2): 69.

Smircich, L. (1983). Concepts of culture and organisational analysis. *Administrative Science Quarterly* 28 (3): 339–358.

Smith, T.D., Eldridge, F., and DeJoy, D.M. (2016). Safety-specific transformational and passive leadership influences on firefighter safety climate perceptions and safety behavior outcomes. *Safety Science* 86: 92–97.

Smith-Jentsch, K.A., Mathieu, J.E., and Kraiger, K. (2005). Investigating linear and interactive effects of shared mental models on safety and efficiency in a field setting. *Journal of Applied Psychology* 90 (3): 523–535.

Snashall, D. (2005). Occupational health in the construction industry. *Scandinavian Journal of Work, Environment & Health* 31: 5–10.

Sobeih, T.M., Salem, O., Daraiseh, N. et al. (2006). Psychosocial factors and musculoskeletal disorders in the construction industry: a systematic review. *Theoretical Issues in Ergonomics Science* 7 (3): 329–344.

Sobeih, T., Salem, O., Genaidy, A. et al. (2009). Psychosocial factors and musculoskeletal disorders in the construction industry. *Journal of Construction Engineering and Management* 135 (4): 267–277.

Song, L., Mohamed, Y., and Abourizk, S.M. (2009). Early contractor involvement in design and its impact on construction schedule performance. *Journal of Management in Engineering* 25 (1): 12–20.

Sorensen, J.N. (2002). Safety culture: a survey of state of the art. *Reliability Engineering and System Safety* 76 (2): 189–204.

Sorensen, G., Barbeau, E.M., Stoddard, A.M. et al. (2007). Tools for health: the efficacy of a tailored intervention targeted for construction laborers. *Cancer Causes & Control* 18 (1): 51–59.

Sorensen, G., Landsbergis, P., Hammer, L. et al. (2011). Preventing chronic disease in the workplace: a workshop report and recommendations. *American Journal of Public Health* 101: S196–S207.

Spangenberg, S., Baarts, C., Dyreborg, J. et al. (2003). Factors contributing to the differences in work related injury rates between Danish and Swedish construction workers. *Safety Science* 41 (6): 517–530.

Sparer, E.H. and Dennerlein, J.T. (2013). Determining safety inspection thresholds for employee incentive programs on construction sites. *Safety Science* 51: 77–84.

Spee, T., Van Duivenbooden, C., and Terwoert, J. (2006). Epoxy resins in the construction industry. *Annals of the New York Academy of Sciences* 1076 (1): 429–438.

Sperling, L.M., Charles, M.B., Ryan, R.A., and Brown, K.A. (2008). Driving Safety: Enhancing Communication Between Clients, Constructors and Designers. In *Proceedings Third International Conference of the Cooperative Research Centre (CRC) for Construction Innovation – Clients Driving Innovation: Benefiting from Innovation*, 12-14 March 2008, Gold Coast, Australia. https://eprints.qut.edu.au/15287/1/15287.pdf (accessed 13 December 2018).

Spiegel, K., Knutson, K., Leproult, R. et al. (2005). Sleep loss: a novel risk factor for insulin resistance and type 2 diabetes. *Journal of Applied Physiology* 99 (5): 2008–2019.

Stagl, K.C., Salas, E., Rosen, M.A. et al. (2007). Distributed team performance: a multi-level review of distributed demography, and decision making. In: *Multi-Level Issues in Organizations and Time* (ed. F. Dansereau and F.J. Yammarino), 11–58. Oxford: JAI Press.

Stattin, M. and Järvholm, B. (2005). Occupation, work environment, and disability pension: a prospective study of construction workers. *Scandinavian Journal of Public Health* 33 (2): 84–90.

Stocks, S.J., McNamee, R., Carder, M., and Agius, R.M. (2010). The incidence of medically reported work-related ill health in the UK construction industry. *Occupational and Environmental Medicine* 67 (8): 574–576.

Stocks, S.J., Turner, S., McNamee, R. et al. (2011). Occupational and work-related ill-health in UK construction workers. *Occupational Medicine* 61 (6): 407–415.

Stricoff, R.S. (2000). Safety performance measurement: identifying prospective indicators with high validity. *Professional Safety* 45: 36–39.

Succar, B. (2009). Building information modelling framework: A research and delivery foundation for industry stakeholders. *Automation in Construction*, 18 (3): 357–37.

Suraji, A., Duff, A.R., and Peckitt, S.J. (2001). Development of a causal model of construction accident causation. *Journal of Construction Engineering and Management* 127: 337–345.

Suter, A.H. (2002). Construction noise: exposure, effects, and the potential for remediation; a review and analysis. *AIHA Journal* 63 (6): 768–789.

Swuste, P., Frijters, A., and Guldenmund, F. (2012). Is it possible to influence safety in the building sector? A literature review extending from 1980 until the present. *Safety Science* 50 (5): 1333–1343.

Szymberski, R. (1997). Construction project safety planning. *TAPPI Journal* 80 (11): 69–74.

Tamburro, N. and Wood, P. (2014). Alliancing in Australia: competing for thought leadership. *Proceedings of the Institution of Civil Engineers – Management, Procurement and Law* 167 (2): 75–82.

Tharaldsen, J.E., Olsen, E., and Rundmo, T. (2008). A longitudinal study of safety climate on the Norwegian continental shelf. *Safety Science* 46 (3): 427–439.

The Construction (Design and Management) Regulations 2007 (UK), (2007) available from www.legislation.gov.uk/uksi/2007/320/contents/made (accessed 27 November 2018).

The Construction (Design and Management) Regulations 2015 (UK), (2015) available from www.legislation.gov.uk/uksi/2015/51/contents/made (accessed 27 November 2018).

Toole, T.M. (2005). Increasing engineers' role in construction safety: opportunities and barriers. *Journal of Professional Issues in Engineering Education and Practice* 131 (3): 199–207.

Toole, T.M. (2007). Design engineers' responses to safety situations. ASCE. *Journal of Professional Issues in Engineering Education and Practice* 133 (2): 126–131.

Törner, M. and Pousette, A. (2009). Safety in construction – a comprehensive description of the characteristics of high safety standards in construction work, from the combined perspective of supervisors and experienced workers. *Journal of Safety Research* 40 (6): 399–409.

Tüchsen, F., Hannerz, H., and Spangenberg, S. (2005). Mortality and morbidity among bridge and tunnel construction workers who worked long hours and long days constructing the Great Belt Fixed Link. *Scandinavian Journal of Work, Environment & Health* 31: 22–26.

Turner, N. and Gray, G.C. (2009). Socially constructing safety. *Human Relations* 62 (9): 1259–1266.

Turner, M. and Lingard, H. (2016a). Improving workers' health in project-based work: job security considerations. *International Journal of Managing Projects in Business* 9 (3): 606–623.

Turner, M., Lingard, H., and Francis, V. (2009). Work-life balance: an exploratory study of supports and barriers in a construction project. *International Journal of Managing Projects in Business* 2 (1): 94–111.

Underhill, E. and Quinlan, M. (2011). How precarious employment affects health and safety at work: the case of temporary agency workers. *Relations Industrielles* 66 (3): 397–421.

Van Broekhuizen, P., van Broekhuizen, F., Cornelissen, R., and Reijnders, L. (2011). Use of nanomaterials in the European construction industry and some occupational health aspects thereof. *Journal of Nanoparticle Research* 13 (2): 447–462.

Van den Berg, T.I., Elders, L.A., and Burdorf, A. (2010). Influence of health and work on early retirement. *Journal of Occupational and Environmental Medicine* 52 (6): 576–583.

Van der Hulst, M. (2003). Long workhours and health. *Scandinavian Journal of Work, Environment and Health* 29 (3): 171–188.

Van Steenbergen, E.F. and Ellemers, N. (2009). Is managing the work–family interface worthwhile? Benefits for employee health and performance. *Journal of Organizational Behavior* 30 (5): 617–642.

Varonen, U. and Mattila, M. (2000). The safety climate and its relationship to safety practices, safety of the work environment and occupational accidents in eight wood-processing companies. *Accident Analysis and Prevention* 32 (6): 761–769.

Vink, P., Urlings, I.J., and van der Molen, H.F. (1997). A participatory ergonomics approach to redesign work of scaffolders. *Safety Science* 26 (1): 75–85.

Vink, P., Miedema, M., Koningsveld, E., and van der Molen, H. (2002). Physical effects of new devices for bricklayers. *International Journal of Occupational Safety and Ergonomics* 8 (1): 71–82.

Vink, P., Koningsveld, E.A., and Molenbroek, J.F. (2006). Positive outcomes of participatory ergonomics in terms of greater comfort and higher productivity. *Applied Ergonomics* 37 (4): 537–546.

Von der Heyde, A., Brandhorst, S., and Kluge, A. (2015). The impact of safety audit timing and framing of the production outcomes on safety-related rule violations in a simulated production environment. *Safety Science* 77: 205–213.

Votano, S. and Sunindijo, R. (2014). Client safety roles in small and medium construction projects in Australia. *Journal of Construction Engineering and Management* 140 (9): 04014045.

Voydanoff, P. (2007). *Work, Family, and Community: Exploring Interconnections*. Mahwah, NJ: Lawrence Erlbaum Associates.

Wachter, J.K. and Yorio, P.L. (2014). A system of safety management practices and worker engagement for reducing and preventing accidents: an empirical and theoretical investigation. *Accident Analysis & Prevention* 68: 117–130.

Waddell, G. and Burton, A.K. (2006). *Is Work Good for Your Health and Well-Being?* London: TSO.

Waddick, P. (2010). Safety culture among subcontractors in the domestic housing construction industry. *Structural Survey* 28 (2): 108–120.

Wahlbeck, K. (2015). Public mental health: the time is ripe for translation of evidence into practice. *World Psychiatry* 14 (1): 36–42.

Wallace, J.C., Popp, E., and Mondore, S. (2006). Safety climate as a mediator between foundation climates and occupational accidents: a group-level investigation. *Journal of Applied Psychology* 91 (3): 681–688.

Warburton, C. (2016). Laing O'Rourke defends 'safety differently' after scathing attack by union. https://www.healthandsafetyatwork.com/accident-reduction/laing-orourke-ucatt (accessed 13 December 2018).

Ware, J.E. (1999). SF-36 health survey in Maruish. In: *The Use of Psychological Testing for Treatment Planning and Outcomes Assessment*, 2e (ed. E. Mark). Mahwah: Lawrence Erlbaum Associates.

Waring, A.E. (1992). Organisational Culture, Management, and Safety. Paper presented at the British Academy of Management 6th Annual Conference, Bradford University, 14–16 September.

Weichbrodt, J. (2015). Safety rules as instruments for organizational control, coordination and knowledge: implications for rules management. *Safety Science* 80: 221–232.

Weick, K.E. (1987). Organizational culture as a source of high reliability. *California Management Review* 29: 112–127.

Weick, K.E., Sutcliffe, K.M., and Obstfeld, D. (1999). Organizing for reliability: processes of collective mindfulness. *Research in Organisational Behavior* 21: 81–123.

Welch, L.S. (2009). Improving work ability in construction workers – let's get to work. *Scandinavian Journal of Work, Environment & Health* 35 (5): 321–324.

Welch, L.S., Michaels, D., and Zoloth, S. (1991). Asbestos-related disease among sheet-metal workers. *Annals of the New York Academy of Sciences* 643 (1): 287–295.

Welch, L.S., Haile, E., Boden, L.I., and Hunting, K.L. (2009). Musculoskeletal disorders among construction roofers — physical function and disability. *Scandinavian Journal of Work, Environment & Health* 35: 56–63.

Welch, L.S., Haile, E., Boden, L.I., and Hunting, K.L. (2010). Impact of musculoskeletal and medical conditions on disability retirement — a longitudinal study among construction roofers. *American Journal of Industrial Medicine* 53 (6): 552–560.

Whyte, J.K., Ewensteinn, B., Hales, M., and Tidd, J. (2007). Visual practices and the objects used in design. *Building Research & Information* 35 (1): 18–27.

Wiegmann, D.A., Zhang, H., von Thanden, T.L. et al. (2004). Safety culture: an integrative review. *The International Journal of Aviation Psychology* 14 (2): 117–134.

Williams, Q., Ochsner, M., Marshall, E. et al. (2010). The impact of a peer-led participatory health and safety training program for Latino day laborers in construction. *Journal of Safety Research* 41 (3): 253–261.

Winkler, C. (2006). Client/contractor relationships in managing health and safety on projects (Research Report 462). http://www.hse.gov.uk/research/rrpdf/rr462.pdf (accessed 13 December 2018).

Workplace Health and Safety Queensland (2013). Silica – Technical guide to managing exposure in the workplace. https://www.worksafe.qld.gov.au/__data/assets/pdf_file/0008/83186/silica_managing_workplace.pdf (accessed 6 February 2018).

WorkSafe Victoria (2005). *Designing Safer Buildings and Structures: A Guide to Section 28 of the Occupational Health and Safety Act 2004*. Melbourne: Victoria State Government.

WorkSafe Victoria. (2016). Preventing and managing work-related stress: A guidebook For employers. https://prod.wsvdigital.com.au/sites/default/files/2018-06/ISBN-Preventing-and-managing-work-related-stress-guidebook-2017-06.pdf (accessed 13 December 2018).

World Health Organization (2008a). *Closing the Gap in a Generation: Health Equity through Action on the Social Determinants of Health*. Geneva: World Health Organization.

World Health Organization (2008b). *Preventing Noncommunicable Diseases in the Workplace through Diet and Physical Activity*. Geneva: World Health Organization.

World Health Organization (2014). *Preventing Suicide: A Global Imperative*. Geneva: World Health Organization.

World Health Organization (2016). *Suicide Rates: Data by Country in 2012*. Geneva: World Health Organization.

Wreathall, J. (2009). Leading? Lagging? Whatever! *Safety Science* 47: 493–494.

Wright, M., Bendig, M., Pavitt, T., and Gibb, A. (2003). The case for CDM: better safer design — a pilot study (Research Report 148). http://www.hse.gov.uk/research/rrpdf/rr148.pdf (accessed 13 December 2018).

Wu, C., Fang, D., and Li, N. (2015). Roles of owners' leadership in construction safety: the case of high-speed railway construction projects in China. *International Journal of Project Management* 33 (8): 1665–1679.

Wu, C., Wang, F., Zou, P.X., and Fang, D. (2016). How safety leadership works among owners, contractors and subcontractors in construction projects. *International Journal of Project Management* 34 (5): 789–805.

Yates, J.K. and Battersby, L.C. (2003). Master builder project delivery system and designer construction knowledge. *Journal of Construction Engineering and Management* 129 (6): 635–644.

Yip, B. and Rowlinson, S. (2006). Coping strategies among construction professionals: cognitive and behavioural efforts to manage job stressors. *Journal for Education in the Built Environment* 1 (2): 70–79.

Zacharatos, A., Barling, J., and Iverson, R.D. (2005). High-performance work systems and occupational safety. *Journal of Applied Psychology* 90 (1): 77–93.

Zhang, R.P., Lingard, H., and Nevin, S. (2015). Development and validation of a multilevel safety climate measurement tool in the construction industry. *Construction Management and Economics* 33 (10): 818–839.

Zhang, R.P., Pirzadeh, P., Lingard, H., and Nevin, S. (2018). Safety climate as a relative concept: exploring variability and change in a dynamic construction project environment. *Engineering, Construction and Architectural Management* 25: 298–316.

Zhou, Q., Fang, D.P., and Wang, X. (2008). A method to identify strategies for the improvement of human safety behavior by considering safety climate and personal experience. *Safety Science* 46: 1406–1419.

Zhou, W., Whyte, J., and Sacks, R. (2012). Construction safety and digital design: a review. *Automation in Construction* 22: 102–111.

Zimmermann, C.L., Cook, T.M., and Rosecrance, J.C. (1997). Operating engineers: work-related musculoskeletal disorders and the trade. *Applied Occupational and Environmental Hygiene* 12 (10): 670–680.

Zohar, D. (1980). Safety climate in industrial organisations: theoretical and applied implications. *Journal of Applied Psychology* 65 (1): 96–102.

Zohar, D. (2000). A group level model of safety climate: testing the effect of group climate on microaccidents in manufacturing jobs. *Journal of Applied Psychology* 85 (4): 587–596.

Zohar, D. (2002a). Modifying supervisory practices to improve subunit safety: a leadership-based intervention model. *Journal of Applied Psychology* 87 (1): 156–163.

Zohar, D. (2002b). The effect of leadership dimensions, safety climate and assigned priorities on minor injuries in work groups. *Journal of Organisational Behavior* 23 (1): 75–92.

Zohar, D. (2008). Safety climate and beyond: a multi-level multi-climate framework. *Safety Science* 46 (3): 376–387.

Zohar, D. and Luria, G. (2004). Climate as a social-cognitive construction of supervisory safety practices: scripts as proxy of behaviour pattern. *Journal of Applied Psychology* 89: 322–333.

Zohar, D. and Tenne-Gazit, O. (2008). Transformational leadership and group interaction as climate antecedents: a social network analysis. *Journal of Applied Psychology* 93 (4): 744–757.

Zorba, E., Karpouzis, A., Zorbas, A. et al. (2013). Occupational dermatoses by type of work in Greece. *Safety and Health at Work* 4 (3): 142–148.

Zwetsloot, G.I.J.M., van Scheppingen, A.R., Bos, E.H. et al. (2013a). The core values that support health, safety, and well-being at work. *Safety and Health at Work* 4 (4): 187–196.

Zwetsloot, G.I., Aaltonen, M., Wybo, J.L. et al. (2013b). The case for research into the zero accident vision. *Safety Science* 58: 41–48.

Index

Note: page numbers to figures are in italics, and those referring to tables in bold.

Integrating Work Health and Safety into Construction Project Management, First Edition.
Helen Lingard and Ron Wakefield.
© 2019 John Wiley & Sons Ltd. Published 2019 by John Wiley & Sons Ltd.